PREPPER COMMUNICATIONS THE EASY WAY

By: Craig E. "Buck," K4IA

ABOUT THE AUTHOR: "Buck," as known on the air, received his first amateur radio license in the mid-sixties as a young teenager. Today, he holds an Amateur Extra Class Radio License.

Buck is also an active instructor and a Volunteer Examiner. The Rappahannock Valley Amateur Radio Club named him the Elmer (Trainer) of the Year three times.

Email: k4ia@EasyWayHamBooks.com

Published by EasyWayHamBooks
130 Caroline St. Fredericksburg, Virginia 22401

This, and other Easy Way books by Craig Buck, are available at Ham Radio Outlet stores and from Amazon:

How to Get on HF – The Easy Way
How to Chase, Work & Confirm DX
Pass Your Amateur Radio Technician Class Test
Pass Your Amateur Radio General Class Test
Pass Your Amateur Radio Extra Class Test
Pass Your General Radiotelephone Operators Test (GROL)

ISBN 978-1724218094

PREPPER COMMUNICATIONS THE EASY WAY

TABLE OF CONTENTS

INTRODUCTION

I have taught many Amateur Radio classes and found that most folks coming into amateur radio these days are preppers. That is different from when I started. We were electronics and science geeks.

Preppers want to communicate, not learn radio theory. However, if you don't know the theory, you can fall prey to outrageous and outlandish advertising claims, misinformation, and rumor, any one of which could be disastrous. This book is going to teach you the theory and practical application of radio.

I will show you realistic expectations for various radio services, how to choose and assemble the equipment you need, and how to fix things when they don't work. This book teaches you The Easy Way to communicate when all else fails.

There are opinions on these pages, and I have avoided qualifying words such as "usually, probably and mostly." Radio is an art as well as a science, and there are exceptions and qualifiers for every rule. I discovered a long time ago, "If you ask five Hams a question, you will get seven different opinions and maybe a fistfight."

I am not an electrical engineer – I flunked math and went to law school. But, I have been a Ham Radio operator long enough to offer you insight into the world of communication.

Please send your comments, corrections, and chastisements to K4ia@EasyWayHamBooks.com. This book prints on-demand, and corrections can appear within 24 hours.

COMMUNICATION FOR PREPPERS

I don't need to convince you of the need for disaster preparation. You've already considered the possibilities and bought this book. Some preppers are satisfied with easing the inconvenience of a few days without power. Others are ready for years of travail.

This book assumes no cell service, no phones, no internet, and no power. That drastic scenario is realistic enough for all levels of preparation.

Wherever you fall on the spectrum, communication is critical. Imagine being hunkered in your bunker and wondering, "Does New York City still exist?" Sobering thought. At the most basic level, you need to hear what is happening in the world around you. That is information you may need to survive. There is comfort in knowing. Even bad news is useful.

How will you contact others for aid and support? Don't limit your horizon to how loud you can yell. Radio is your source of information and medium of communication. This book is about radio, and we'll start by exploring some theory to help you understand what is to follow.

The wise store up choice food and olive oil, but fools gulp their's down. Psalms 21:20

FREQUENCY

We'll begin at the beginning with the concept of frequency. Electricity can travel by direct current or alternating current. Direct current flows in only one direction. A battery provides direct current.

Frequency Alternating current reverses direction, and "frequency" is the term describing the number of times per second that an alternating current makes a complete cycle. We measure frequency in cycles per second. The term "hertz" is a shortcut for "cycles per second." You tune a radio by changing the frequency.

The power in a wall socket reverses 60 times a second and is, therefore, 60 hertz. A radio signal may be anywhere from a few thousand hertz up to millions of hertz. WMAL, a local AM radio station, broadcasts on a frequency of 630,000 Hz or 630 kilohertz, abbreviated 630 kHz, and standing for 630 thousand Hertz. Citizens Band radios operate around 27 MHz, 27 megahertz, or 27 million hertz.

Electromagnetic energy A radio wave is electromagnetic energy. The reversing current causes an electromagnetic field to form around the antenna wire. That field radiates (spreads) and induces a current into the receiving antenna, even if that antenna is thousands of miles away. That is how radio works.

I can't see it, so to me, it is just magic. A signal leaves my wire and is heard on the other side of the world. The signal may only be millionths of a volt, but our receiver circuits are sensitive enough to pick it up. To quote Samuel Morse's first message, "What hath God wrought?"

Electromagnetic spectrum We divide the electromagnetic spectrum into three categories:

- 3 to 30 MHz (HF High Frequency)
- 30 to 300 MHz (VHF Very High Frequency)
- 300 to 3000 MHz (UHF Ultra High Frequency)

Wavelength A radio wave travels through free space at the speed of light, approximately 300,000,000 meters per second. If we know the wave is traveling at 300 million meters per second, and the frequency is 144 million cycles per second, we can calculate how far the wave will travel in one cycle. Wavelength is the name for the distance a radio wave travels during one complete cycle.

The magic formula is to divide 300 by the frequency in MHz to get the length in meters. Or, divide 300 by the wavelength in meters to get MHz.

The higher the frequency, the more often the wave reverses direction, and the less distance it can travel in a cycle. Therefore, the wavelength gets shorter as the frequency increases. Knowing the wavelength helps us calculate the optimal length for an antenna.

Band Now, consider the concept of a "band." A band is a group of frequencies, and referring to the band is a shortcut. You are familiar with the AM radio band that stretches from 550 kHz to 1600 kHz. (Kilohertz)

Bands are often associated with the wavelength. The two-meter amateur band goes from 144 – 148 MHz.

We'll learn more later about the characteristics of different frequencies and why you would pick one group of frequencies (band) over another to accomplish your communication goal.

MODES

The mode is the transmission type. You know about AM and FM modes from your car radio. They are different means of modulating and receiving radio signals. Modulation describes combining speech with a radio frequency carrier. The RF carrier is the base frequency of the transmission. AM is amplitude modulation, and FM is frequency modulation.

FM MODE

FM mode (frequency modulation), communicates information by jiggling the frequency of the transmission resulting in a signal which is relatively wide but clear and high fidelity. FM mode is used mainly on the VHF and UHF bands.

AM MODE

The AM Mode (amplitude modulation) communicates by jiggling the amplitude of the radio wave. The audio information varies the power of the wave, and the result is the center frequency called the carrier and a sideband on either side of the carrier varying in amplitude in accord with the audio. Shortwave broadcast stations use AM.

SINGLE SIDEBAND MODE (SSB)

Single sideband (SSB) is a form of amplitude modulation (AM). Look at the diagram. The bottom

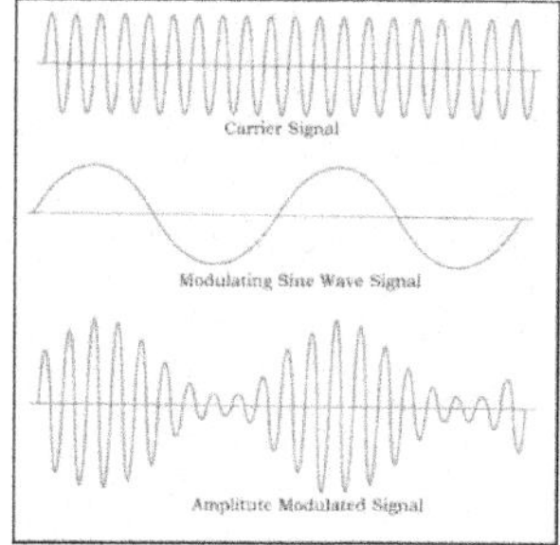

waveform is an AM signal. To make SSB, filters in the transmitter strip away the carrier and one of the sidebands leaving only a single sideband. The remaining sideband could be either on the upper or lower side of the carrier. You will hear this referred to as upper sideband or lower sideband. By convention, frequencies above 14 MHz use upper sideband (USB).

Single sideband sounds like Donald Duck until properly tuned. However, SSB is a superior voice mode because it requires less power and takes up less bandwidth (space) than either AM or FM. Amateur Radio operators use SSB as the primary voice mode.

PACKET MODE

Packet mode uses bursts of information (data). These packets are decoded and checked at the receiving end and, if there are errors, the receiving station automatically requests a repeat. Packet is very effective as a data mode to transfer messages or lists of information.

CW MODE (MORSE CODE)

CW, which stands for continuous wave or carrier wave, is the most basic of digital modes. The carrier is turned on and off, making the dots and dashes we recognize as Morse code.

In comparison to AM/FM/SSB, CW has the narrowest bandwidth and has the best chance of being understood under the worst conditions.

There is a chapter on learning Morse code later in the book. Morse code is not required for any level of amateur radio license. Morse is still popular among hams.

POWER SOURCES

How do you power your radio? If the mains power is out, how much battery backup do you need? Understanding a bit of electronic theory sorts it out.

VOLTS, AMPERES, AND RESISTANCE

Voltage represents electromotive force. Think of Voltage as pressure in the line. All that force can sit there as it does with an electrical outlet. With nothing plugged in, the 120 volts is there, but not doing anything.

Amperes represent the current or amount of electricity flowing through the circuit. When you turn on a light, current flows through the wire to power the bulb. Now, that voltage is pushing the current through the load (light bulb).

Resistance, or load, also affects the flow. With more load or resistance, less current flows.

Power is the amount of work done by the pressure and flow.

You can intuitively feel there is a relation between all four:
More pressure = more current flow.
More resistance = less current flow.
More current or more pressure = more power.
Are these mathematically related? Georg Ohm thought so and came up with his famous formula known as Ohm's Law.

Ohm's Law: Voltage = Amperes times Resistance.
For this amazing feat, he named the unit of resistance after himself (Ohm).

E symbolizes voltage (Electromotive Force).
I is Amps or Amperage (flow).
R is Resistance.

So the formula for Ohm's Law looks like E=IR.
If you know two of the variables, you can calculate the third. The easy way is to use the magic circle and draw it as the Eagle flies over the Indian and the Rock.

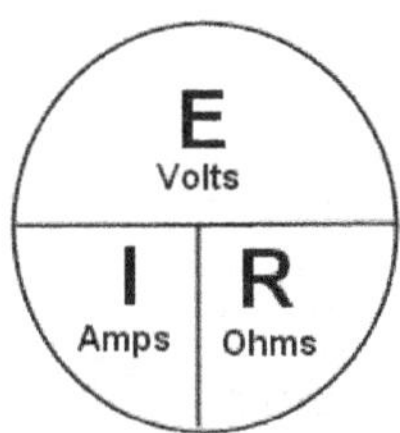

Put your thumb on the value you are solving for, and the formula is what's left. For example, if you are solving for R, cover up the R, and the answer is E/I, voltage divided by amperage. If you are solving for E, cover it up, and the answer is I x R, amps times ohms.

POWER IN WATTS

Watts is a measure of electrical power or work.

The formula to calculate electrical power in a DC circuit is Power equals Voltage (E) multiplied by Current (I). It is easy as PIE, P=IE

We have another magic circle to help us.

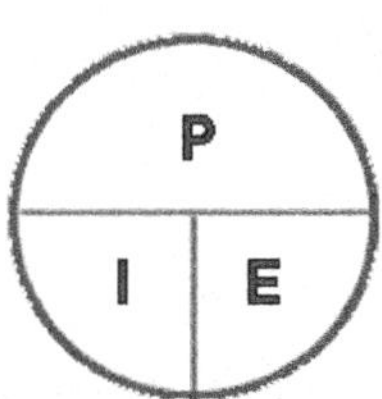

POWER SOURCES

If we know a particular piece of equipment uses 100 watts at 120 volts, we know it draws about 100/120 or .8 amperes (amps). If it uses 100 watts at 12 volts, it will draw 100/12 or 8 amps. This information tells you what you will need from a generator or battery.

GENERATORS

A generator may be overkill just to run a radio but can power the rest of your household. The downside is that generators are noisy, distracting, generate poisonous carbon monoxide gas, and require a source of clean fuel.

Construction Generators I refer to the gas or diesel-powered models on wheels as construction generators. They provide 4,000 to 6,000 watts of power. The rating refers to peak power, and the sustained power rating is less. Add the power consumption for all the appliances you intend to run at the same time and choose a generator that will handle that load. Don't consider just the running wattage of the equipment. The starting load can be much higher. Pumps, motors, and compressors can require seven times the operating power to start. If a small air conditioner draws 1,000 watts running, it might need 7,000 watts to get going.

On the other hand, you may not need as much power as you think because equipment like a refrigerator, freezer, and other appliances won't be drawing current at the same time.

The problems with construction generators are:

- big, heavy, hard to move
- loud
- run at a constant speed no matter what the load and consume a lot of fuel
- may not produce clean, well-regulated power suitable for electronics.

How much fuel the generator consumes depends on the load. A heavier load burns more. One gallon of gas will last about 3 hours on a 50% load. Plan on 12 – 15 gallons a day for heavier loads. You will have to store and stabilize a lot of fuel.

The Uniform Fire Code prohibits storing more than 5 gallons of gasoline in residential areas. You must store gasoline in an open place where fumes can dissipate. Gas vapor is heavier than air, and explosive amounts can accumulate in a low spot.

Running generators must be well ventilated. Deadly carbon monoxide is a by-product of internal combustion engines. It is colorless and odorless, so you will have no warning before you pass out. Do not use generators indoors, in enclosed areas, or near air intake vents. If you see someone with bright blue lips or fingernails, suspect CO poisoning and get them to fresh air, oxygen and medical attention immediately.

Some construction generators are not well regulated, meaning the voltage and frequency vary. They do not produce a clean alternating current wave. They are suitable for power tools and lights, but electronics like radios and computers might not work as expected.

Inverter Generators The newer "inverter" generators have several advantages:

- much smaller, lighter, and quieter
- speed changes to match the load, so they consume less fuel
- power output is well-regulated and a pure sine wave your electronics will appreciate.

I have a Honda 2000-watt inverter generator. The Honda is light enough to carry and so quiet you can hold a conversation next to it. The fuel needed is 2 – 6 gallons per day, depending on the load.

While the 2000-watt rating seems low compared to the construction generators, this may be all you need. It is certainly enough for radios. You can fudge a bit with other equipment because things like a refrigerator or freezer don't run continuously. If you keep the door shut, they will stay cold for hours. When we lost power during a hurricane, I was able to keep going by cycling the appliances, including a small room air-conditioner. This kept the beer cold and the TV on for three days using less than 5 gallons of gas.

Inverter generators are more expensive, but they would be my recommendation.

Stay away from the cheap 2-cycle generators sold at places like Harbor Freight. They are very noisy, both audibly and electrically, and have terrible voltage regulation.

BATTERIES

You can rely on household power for your radio with a battery backup system when the mains power fails. Combine that with a solar panel to recharge the battery. How much solar panel and how much battery depends on what you plan to run.

Battery Types A supply of non-rechargeable alkaline batteries will store for years and run your low-power portable devices.

Rechargeables are better for more robust and sustained use. Rechargeable batteries include:

- Nickel Cadmium (NiCad)
- Nickel-metal hydride (NiMH)
- Lithium-ion (Li-Ion)
- Lead-acid.

NiCads have fallen out of favor because they suffer from a memory effect, have a high self-discharge rate[1], and are heavy compared to their energy capacity. If you leave them charged at 50% for too long, or discharge them to 50% and then recharge, the chemistry changes and the battery will not charge more than 50%. You only have half a battery.

NiMH NiMH batteries provide a nominal 1.2 volts per cell, slightly less than alkaline (1.5 volts). If you put four in a radio, they total 4.8 volts, not the full 6 volts. The 20% reduced voltage will reduce the power output of a transmitter by 20% as well. That is why some handheld-radio battery holders fit five cells, to bring the voltage back up to 6 volts. NiMH batteries require a special charger circuit to prevent over-charging.

Li-Ion Li-Ion batteries, and their cousins, the LiPo (Lithium Polymer), have similar characteristics to NiMH, but they recharge faster, are smaller and lighter, and hold more energy per pound than other batteries. Hybrid cars and laptop computers use Li-Ion batteries. They require a special charging circuit and provide 3.6 volts per cell. Li-Ion batteries require no maintenance, are one-third the weight, and one-half the volume of lead-acid. They last in service up to five times longer. Li-Ion maintains its voltage through most of the discharge cycle whereas other batteries experience a gradual voltage drop.

A Li-Ion battery can be discharged 80% and still recover. A lead-acid battery dies at 50%. Newer cars are arriving with Li-Ion starter batteries, and they are a good, but expensive, replacement for lead-acid.

Lead Acid Lead-acid batteries are very heavy and designed to crank a car engine. They produce a large amount of power in a short burst. If you use lead-acid

---

[1] They lose energy even when not being used.

batteries for sustained applications, get a deep-cycle trolling motor or golf cart battery. Their design allows them to produce power over a longer time.

If the power is out, recharge a 12-volt battery by connecting it in parallel (positive to positive and negative to negative) with a vehicle's battery and idle the engine. A full tank of gas should run 72 hours and, charging a few hours a day, it would last for weeks. Consider your car as a source of energy in a power outage. Heed the warnings about carbon monoxide poisoning and don't do this in a garage.

Solar Power You can also recharge batteries using solar power. Be wary of the power capacity claims made by solar sellers. Their figures assume bright sun, perfect alignment, a temperature below 75F, a very clean panel, and a bit of imagination. You will need more panels and more time to recharge than you think.

Solar chargers contain special circuits to regulate the charge cycle, and you must make sure it matches your battery type. Using a lead-acid charger on a Li-Ion battery would be a mistake.

Battery Dangers If you charge or discharge any battery too quickly, it can overheat and give off flammable gas or explode. Another safety hazard is that shorting the terminals can cause burns, fire, or an explosion. Batteries deliver a tremendous amount of current that can melt a ring or tool. Take your ring or metal watchband off and keep your tools clear when working around batteries.

Battery Capacity Batteries ratings are in amp-hours. "Amp" is an abbreviation for ampere, the amount of current flow in a circuit. You would think a battery rated at 4 amp-hours should be able to produce 1 amp for 4 hours or 2 amps for 2 hours. It doesn't work out that way because the battery provides less power as it discharges and won't last as long if discharged quickly. Battery capacity is based on a 20-hour discharge rate. Your 4 amp-hour battery will supply 1/5 amp for 20 hours. That is enough for a handheld transceiver and not much else.

Also, the rating assumes you discharge the battery completely, but you should never do that because it would permanently damage the battery. A good rule of thumb is half the stated rating for lead-acid and 75% for Li-Ion. Your 7 amp-hour lead-acid battery is only usable for 3.5 amp-hours of service. Use the amp-hour rating to plan and compare batteries.

If your transceiver outputs 50 watts, it requires 85 watts input.[2] That will translate to around 6.5 amps at 13.8 volts. A 7 amp-hour battery de-rated by 50% would last 30 minutes. The calculation gets complicated when applied to real life. A transmitter might need 85 watts peak when transmitting, but it isn't transmitting all the time. The duty cycle is such that you are mostly listening and might only transmit 10% of the time. Listening might only draw 1 amp, so your 7 amp-hour battery would last much longer.

So, how much battery do you need? The answer depends on how much you want to run and for how long. Bigger is better. As a general rule of thumb, to last through 12 hours of operation, you need, at a minimum, one amp-hour of capacity for each watt of transmitter output. That is the recommendation of the Radio Amateur Civil Radio Service (RACES), a standby radio service established by FEMA and the FCC.

RACES assumes you will be doing a lot of transmitting during your 12-hour shift. Another general answer is you want a lead-acid battery of least 30 amp-hours

---

[2] Assuming the transceiver is 60% efficient.

capacity if you are transmitting at 25 watts and running a computer or other accessories. That battery will weigh 25 pounds.

You can get away with 20 amp-hours from a Li-Ion battery at one-third the weight because it will produce full output down to an 80% discharge. In either case, monitor the voltage, so it does not drop below the critical level for the battery's chemistry.

You can extend the battery's discharge cycle by attaching a solar-panel to replace the energy used. If you take 10 amp-hours out of the battery, you need to put 10 back in to maintain the charge. If the solar panel is rated at 100 watts, that means it will output 7 amps at 14 volts (P=IE) and it will take 10/7 =1 ½ hours to recharge. The actual time will be longer (maybe twice as long) because of inefficiencies in the charge controller circuits and less than ideal sun conditions. Judge your solar panel needs accordingly.

Solar charge controllers have meters designed to measure power in and out.

POWER SUPPLIES

Solid-state transceivers require a nominal 13.8 volts DC. That can come from a battery system, or using a power supply to convert your generator or house voltage (nominal 120 volts, AC).

Volts measure pressure in the line. Your household electric outlet has 120 volts of pressure. Amperes or amps measure the current flow in the circuit. Volts times amps = watts (power).

The current rating of a power supply is the number of output amps it can provide before the voltage sags or the fuse blows. You will hear a power supply referred to as a "25 amp supply."

At 100 watts output, your 60% efficient transceiver amplifier circuit will require 167 watts and draw 167/13.8 = around 12 amps of current from the 13.8-volt source. Add a few amps for overhead such as the other circuits in the radio, dial lights, etc. and you can estimate the supply needs to be at least 20 amps.

Don't scrimp and get a 20-amp power supply for a 100-watt transceiver. It will be marginal and leave you no room for any accessories, such as another radio. The additional cost of a 25-amp, or better a 35-amp supply is not significant, and you won't be sorry. I like volt and amp meters so I can see what is happening. Meters don't add much to the cost, either.

Linear Power Supplies There are two designs for AC power supplies. The so-called "linear" types use a large transformer to convert the 120 volt AC household current to 13.8 volts DC for the radio. Linear supplies are heavy and very rugged. Mine has been on almost continuously for 15 years without a glitch.

Switcher Power Supplies Switcher type power supplies are much lighter and smaller. They convert the house mains 60 hertz AC to a higher frequency, which allows using much smaller and lighter transformers to reduce the voltage. The frequency conversion can generate radio frequency interference (RFI), and switcher supplies have gotten a bad rap on that account.

Also, the switcher circuits are one more thing to fail. Check the reviews before you buy and pay particular attention to complaints about RF hash. I have used both types with no problems. Consider a switcher if you will need to move the power supply often or carry it while traveling.

POWER SOURCES

Keep all types of power supplies away from the transceiver to reduce coupling of noise or hum. Setting them close together is asking for trouble. My supply is on the floor under the desk.

Power Cord If you buy a new transceiver, it will come with a sufficiently stout power supply cable and the proper plug to fit the radio. A used transceiver may have neither. Don't scrimp with cheap speaker wire to power your rig.

Resistance in the wire makes the voltage drop. If your transceiver draws twenty amps, twelve feet of number 16-gauge speaker wire will decrease the supply voltage by two volts at the radio. Low voltage starves your transmitter of the power it needs, and sagging power will distort your signal or shut down the rig.

Twelve feet of 10-gauge[3] wire will only drop about half a volt. Use 8 or 10-gauge wire designated as power cord. If you use 12-gauge, keep the length to six feet or shorter. The right supply cord will save you many headaches.

Watch your polarity! Attaching the power cord backward can do severe damage to your radio. Power cords have one red and one black wire. Red is positive. Black is negative. Both wires should have fuses. Anderson Power Poles are a standard plug-in connector that have one red and one black side, making it harder to reverse polarity.

---

[3] Lower gauge numbers mean thicker wire.

PROPAGATION

Propagation is the term used to describe the spread of radio signals. The greatest influence is from the ionosphere, which is clouds and layers of charged particles above the earth. Propagation changes with the seasons, sunspot cycles, and during the day or night. Here are some general guidelines.

UHF and VHF signals (30 MHz and above) do not bounce off the ionosphere. They are used to communicate with satellites because the signals pass through the ionosphere.[4]

UHF and VHF frequencies can bounce off buildings and obstructions, and if they bounce off multiple surfaces, you will experience multi-path distortion or echoes. That would result in a fluttery or distorted signal. Moving will get out of the echo chamber. UHF and VHF frequencies are line-of-sight and best suited for local communications.

HF signals (3 – 30 MHz) can bounce off the ionosphere, but the various frequency bands react differently. The general rule is higher HF frequencies are open during the day and periods of high solar activity. The lower frequencies are open at night. The twenty-meter Amateur band (14 MHz) is in the middle and can be open somewhere 24 hours a day. As you progress above and below twenty meters, the general rule is more noticeable.

If you want to establish a long-distance link, you need a ham license to operate on HF.

---

[4] VHF signals can benefit from atmospheric conditions called tropospheric ducting but that is not predictable or reliable for our purposes.

PROPAGATION

Here is a guideline for Amateur and Shortwave Radio bands and propagation:

Band in Meters	**Band MHz**	**Characteristics**
160 M	1.8 - 2	Amateur Radio band. Nighttime only.
120 M	2.3-2.495	Local in tropical regions. Time stations at 2.5 MHz.
90 M	3.2–3.4	Local in tropical regions with limited long-distance nighttime reception. Canadian time station CHU is on 3.33 MHz.
75 M	3.6 - 4.0	Amateur Radio voice band. Local during the day and longer at night.
41 M	3.9–4	Mostly Eastern Hemisphere. Shared with the North American Amateur Radio Band.
60 M	4.75-5.06	Local in tropical regions. WWV time station on 5 MHz.
49 M	5.8–6.2	Nighttime band. Poor for long distance during daytime.

41 M	7.2–7.45	Reasonably good nighttime reception especially from Europe. Shares part of the 40 M Amateur Radio band.
40 M	7.125 - 7.3	Amateur Radio voice band. Several hundred miles daytime. Long distances at night.
31 M	9.4–9.9	Most used shortwave band. Good year-round nighttime and best daytime reception is in winter. WWV time station on 10 MHz.
25 M	11.6–12.1	Best during summer and just before and after sunset all year
22 M	13.57–13.87	Similar to the 19 M band. best in summer.
20 M	14.150 – 14.350	Amateur Radio voice band. Long distance reception most of the day and night.
19 M	15.1–15.83	Day reception good, night reception variable; best during summer. Time station WWV on 15 MHz.
17 M	18.110 – 18.168	Amateur Radio voice band. Best during daytime.

16 M	17.48–17.9	Day reception good. Night reception varies seasonally, with summer best.
15 M	18.9–19.02	Lightly used daytime band
15 M	21.2 - 21.450	Amateur Radio voice band. Trans-equatorial contacts common during the day.
13 M	21.45–21.85	Erratic daytime reception, with very little night reception. Requires good solar flux numbers.
12 M	24.930 – 24.990	Amateur Radio voice band. Requires good solar flux.
11 M	25.6–26.1	Seldom used for shortwave broadcasting. Reception is poor in the low solar cycle, but potentially excellent with good solar flux.
11 M	26.965 – 27.405	Citizens Band. Mostly local contacts.
10 M	28.3 - 29.7	Amateur Radio voice band. Requires good solar flux

		numbers for distance reception.

Propagation on the HF bands may be too long for your use. You might hear signals from Europe but nothing closer. Nearby signals are bouncing over your head because you are in what we call the "skip zone." You can overcome this shortcoming by proper choice of antenna.

WHERE DO YOU NEED TO COMMUNICATE?

Decide where you need to communicate and design your station accordingly. If your goal is to communicate with the state capital, 100 miles away, the twenty-meter amateur band (14 MHz) is not the answer. As a ham, I would be on eighty or forty meters using a Near Vertical Incidence Skywave Antenna (NVIS). Using a GMRS transceiver, I would set up directional antennas as high as I could get them on both ends of the circuit. Read more in the Antenna chapter.

The time to solve these dilemmas is now when there is no crisis, no anxiety, no pressure, and no one's life is at stake. Practice and experiment during the easy times to prepare for the tough ones.

Remember the seven P's:
Proper Prior Planning Prevents Pretty Poor Performance.

RECEIVERS AND SCANNERS

You will want to know what is happening – locally, nationally, and internationally. Radio is the way to get news, comfort, and entertainment with no power, no internet, and no phones.

During the Cold War, the CONELRAD radio system used the AM radio band as a national emergency broadcast network. CONELRAD emblems marked the radio dials, and the public was to tune there in the event of an emergency. All other stations went silent to deny enemy bombers radio beacons.

Sadly, this system does not exist today. Our preparedness is worse. Few commercial radio stations have sustainable backup power, and many only broadcast an internet feed. If there is no power or no internet, you won't hear any local stations. They will be off the air.

Long range AM Your local stations may be silent, but at night, you can hear long-range activity on the AM radio band. That is because signals at those frequencies reflect off the ionosphere and can be heard at a distance.[5] It only works at night because another ionospheric layer blocks those frequencies during the day. Tune around after sunset for long-distance AM-band signals. Appendix D is a list of 50,000-watt AM mega-stations.

The AM/FM radio in your car might not be adequate. If you want consistent coverage, you need a better receiver and antenna. Often a long wire attached to the radio's antenna will improve reception.

Shortwave Shortwave operates in the HF (high-frequency) bands between 3 and 30 MHz. At these

---

[5] FM radio is on VHF frequencies that do not bounce off the ionosphere.

frequencies, signals can bounce off the ionosphere during the day or night, traveling hundreds or thousands of miles.

To hear shortwave, you need a "general-coverage" receiver. A radio that receives SSB (Single Sideband) will also be able to hear Amateur Radio operators, and they can be a good source of information.

NOAA Some models also cover the National Oceanic and Atmospheric Administration (NOAA) weather alert frequencies on seven channels between 162.400 MHz to 162.550 MHz. There, you will hear continuous 24/7 bulletins from the National Weather Service. NOAA has Specific Area Message Encoding (SAME) for weather and non-weather emergency messages. If your radio is SAME enabled, key in the six-digit code for your area, and you will hear local emergency alerts break into the regular weather reports.

You should have a quality AM/FM/Shortwave/NOAA radio in your kit. One that is also SAME enabled is even better. They come in all shapes, colors, and sizes. Multiple power sources are helpful, and some will accept batteries, solar power, or a crank to charge an internal rechargeable battery. Avoid toys and get a decent radio, with a good-size speaker and headphones. The ones with frequency synthesizers and digital readouts are better (more precise and stable) than the dial models. Spend at least $100. Get a cheaper model to keep the kids entertained.

Hints for operation:

- For the best shortwave reception, make an extension antenna using an alligator clip and 10 -20 feet of wire. Attach it to the built-in telescoping antenna. The extra length will increase the signal gathering ability. It might also be "too much" and overload the receiver, so experiment.

RECEIVERS AND SCANNERS

- Antennas are directional. Move the antenna around for the best reception. Get it up high. Try hanging it vertically.
- Don't store the radio with batteries installed. They will die and leak, often ruining the radio. Store batteries separately.
- Play with the radio to familiarize yourself with the operation and learn the frequencies and stations you can reasonably expect to hear.

Scanners Scanners got the name because they scan a group of frequencies and stop whenever there is a signal. You program the scanner to check on a range of frequencies. Most HTs and shortwave radios have scanning functions.

Dedicated police scanners were fun before police and emergency services started encrypting transmissions. Now, the scanner will stop on a signal, but you won't be able to understand the transmission.

You can still buy scanners, but you may not be able to hear anything. Even if you had the technology, it is against the law to decrypt encrypted police or fire transmissions. Before you spend money on a dedicated scanner, check one of the many online databases to see if it will be worthwhile in your area. You may find the only transmission you can understand will be airport baggage handlers and race-car pit crews.

UNLICENSED RADIO SERVICES

Listening is all well and good, but what if you want to talk to someone? Now, you need a transceiver, a transmitter and receiver, all in one box. Transmitting is regulated by national and international law. You can buy a radio, and you can listen, but you may not transmit except as the laws allow. We'll start with the simple services that do not require a license to operate.

The first question everyone asks is, "How far will it transmit?" The answer is going to disappoint you. Advertising claims are wildly exaggerated and could only happen under mountain-top to mountain-top conditions, sliding downhill both ways with the wind at your back.

CITIZENS BAND

CB was all the craze in the 1970s, and one might say it suffered from too much of a good thing. CB became popular and degenerated into a brawl. Rude behavior and offensive language have turned most people away. However, there are millions of CB radios in place, and you will hear action in an emergency.

CB radios operate on 40 channels[6] around 27 MHz and are limited to 4 watts output of AM or 12 watts of SSB. For comparison, a night light is 5 watts, so CB is a low power resource. The FCC[7] tests radios entering the marketplace, and they cannot exceed those specifications, no matter what the guy in the CB shop

---

[6] Before 1977, CB was limited to 23 channels. If you see a 23-channel rig for sale, it is probably more than 40 years old.

[7] The Federal Communications Commission regulates and policies communication services in the United States.

tries to tell you. If the CB shop tinkered with the radio to exceed the limits, they have broken the law.

You will see CB amplifiers for sale, commonly referred to as "kickers," "after-burners," "linears," or "pills." Amplifiers and souped-up radios are illegal and probably do not meet proper engineering's spectral purity and filtering requirements. In other words, they can cause interference, splatter, and distortion.

Maybe you don't care if the radio is legal or not. You only intend to use it under circumstances where an illegal radio will be the least of anyone's concerns. I get it, but if the radio has been tampered with, or attached to an amplifier, it is probably out of specifications, meaning you'll have a lousy-sounding signal or unsafe operation. There are lots of people listening, and it is not difficult to triangulate on a signal and find who is causing the interference.

In truth, the joke is on whoever pays money to get the extra power of a souped-up CB radio or amplifier. Line of sight, the curvature of the earth, intervening buildings, hills, and other obstructions all limit the useable distance. Increasing power won't make much difference, if any. It will only make you louder to stations who might hear you anyway.

The other consideration is, you need to take the radio out of the box and use it to become familiar with the operation and develop a realistic assessment of what it can do. You won't do that if you are worried about a visit from an FCC field agent. A bad signal will draw attention.

CB radios operate off 13.8 volts, which is the nominal voltage of a car battery. CBs consume little power and can be plugged into a cigarette lighter. Setting up at home requires a power supply to convert the house voltage to 13.8 volts. If there is no house power, you need a battery.

The most common CB antenna is a vertical; a stick anywhere from a foot or two to 102 inches long. The shorter antennas have a built-in coil that makes them appear electrically longer. Such tricks come at a price, and the short antennas are not as efficient as full-size. Shortened CB antennas often come with a magnetic mount to stick on the car.

Radio waves travel in straight lines, so communication is limited to line-of-sight unless the wave bounces off the ionosphere. That can happen and support long-distance contacts on 27 MHz under the right conditions, but they are rare and unpredictable. Therefore, the reasonable range of a CB radio tops out at 2 – 5 miles, more if you are mountain-top to mountain-top.

You will spend $100 and up for a decent CB radio, plus the cost of an antenna ($15 - $50). Used CBs are cheap and plentiful.

Hints for CB operation:

- Channel 3 is the prepper channel.
- Channel 19 is the "calling channel" where most conversation starts. Switch to another channel once you make contact.
- Channel 9 is the emergency channel, often monitored by police. Stay off except to report an emergency.
- Mount your antenna high for maximum coverage.
- Use SSB for maximum talk power.
- You can power from a cigarette lighter but might have better results if you attach directly to the battery. Connecting to the battery avoids noise from computer circuits and the car's engine.

FAMILY RADIO SERVICE

HTs are handheld devices (Handie Talkies or Walkie Talkies), and several radio services use this format. You can find HTs for Citizens Band, but they are more common in the Family Radio Service (FRS) and General Mobile Radio Service (GMRS) because an effective antenna at CB frequencies (102 inches) is too long to fit on an HT. An antenna for FRS or GMRS frequencies is only 6 inches long.

FRS radios are limited to 2 watts output and operate at a frequency between 462 MHz and 467 MHz in FM mode. Channels 8-14 are limited to ½ watt. FRS radios must have permanently attached antennas so you can't substitute something better.

You can program individual channels with a tone code, so you only hear others using the same tone. That helps sort out interference in crowded conditions, but it doesn't guarantee privacy. Encryption is never allowed in any non-government service.

Hints for FRS operation:

- Channel 1 is designated the calling frequency. Make contact there but move to another channel to continue the conversation.
- Channel 3 is the emergency channel.
- Agree on tone codes for your group to reduce interference.
- Always hold your HT vertically, so everyone maintains vertical polarization. (See polarization in the Antenna chapter).

In my humble opinion, FRS radios are toys. They are crippled purposely to restrict coverage and don't have sufficient range or flexibility to do much more than communicate to the backyard or serve as an intercom. In crowded conditions, interference makes FRS unusable.

GENERAL MOBILE RADIO SERVICE

General Mobile Radio Service (GMRS) requires a license, but I include it in the unlicensed services because the license is only a registration; there is no test. One license covers the whole family and costs $70 for ten years.

You can apply for a license online at www.fcc.gov/uls. First, you register to establish an account with the FCC. That gets you an FRN number, which is your identification for all future business with the FCC. You will also choose a password for your account. Then, you can proceed to the application page for your GMRS license. Within 24 hours, you get an email with a link to your license. My link wouldn't open, so I logged on to the FCC.gov.uls website using my FRN and found my call sign, WRBX217.

You must identify by announcing your call sign in English (or Morse Code) at the end of a series of transmissions and every 15 minutes during a conversation. The full text of the GMRS regulations is in Subpart E, Part 95, Title 47 of the Code of Federal Regulations. Notable is this section on prohibited uses:

§95.1733 Prohibited GMRS uses.

(a) In addition to the prohibited uses outlined in §95.333 of this chapter, GMRS stations must not communicate:

(1) Messages in connection with any activity which is against Federal, State, or local law;

(2) False or deceptive messages;

(3) Coded messages or messages with hidden meanings ("10 codes" are permissible);

(4) Music, whistling, sound effects or material to amuse or entertain;

(5) Advertisements or offers for the sale of goods or services;

(6) Advertisements for a political candidate or political campaign (messages about the campaign business may be communicated);

(7) International distress signals, such as the word "Mayday" (except when on a ship, aircraft or other vehicle in immediate danger to ask for help);

(8) Messages which are both conveyed by a wireline control link and transmitted by a GMRS station;

(9) Messages (except emergency messages) to any station in the Amateur Radio Service, to any unauthorized station, or to any foreign station;

(10) Continuous or uninterrupted transmissions, except for communications involving the immediate safety of life or property; and

(11) Messages for public address systems.

(12) The provision of §95.333 apply, however, if the licensee is a corporation and the license so indicates, it may use its GMRS system to furnish non-profit radio communication service to its parent corporation, to another subsidiary of the same parent, or to its own subsidiary.

(b) GMRS stations must not be used for one-way communications other than those listed in §95.1731(b). Initial transmissions to establish two-way communications and data transmissions listed in §95.1731(d) are not considered to be one-way communications for the purposes of this section.

GMRS shares a few common frequencies with FRS but has a maximum power of 50 watts on some channels vs. 2 watts for FRS. Privacy tones are available, and you can use an external antenna. Fifty watts is quite a bit of power but would only be available from a base/mobile station and not an HT. The breakdown of frequencies and power is in Appendix E.

You may not use GMRS in Canada or north of what is known as "Line A." Line A is an imaginary line along the northern border of the United States, an area including Detroit, Buffalo, Cleveland, and Seattle. There is also a "Line C" along the Alaska/Canada border. If you live or travel in those areas, check a map. https://www.fcc.gov/engineering-technology/frequency-coordination-canada-below-470-mhz#LineA

GMRS is limited to line of sight as there is no reflection off the ionosphere at these frequencies. If you have a clear line of sight, meaning absolutely nothing between the two stations, expect 10 - 30 miles coverage at 50 watts.[8]

GMRS does allow eight channels for repeaters, and they will expand your coverage area, if available. See the chapter on Repeaters.

A package with a 15-watt base/mobile station, magnetic mount antenna, and 2 HTs runs less than $200.

MULTI-USE RADIO SERVICE

Another service in the "private, two-way, short-distance voice or data" category is called Multi-Use Radio Service. MURS is very similar to FRS and GMRS but operates on five channels, around 151 MHz. This frequency is too high for skip off the ionosphere, so it is again line-of-sight.

Power is limited to 2 watts output, and the antenna cannot be more than 60 feet above the ground. On the ground, I wouldn't expect any better coverage than FRS (2-5 miles in the clear). With an elevated antenna, perhaps up to 15 miles.

HINTS FOR HT OPERATION

- With all HTs, use fresh batteries. Low voltage batteries will drastically cut power and cause distortion.
- Don't store the radio with batteries installed. They will die and leak, often ruining the radio. Store batteries separately.
- Don't let rechargeable batteries sit idle. They can self-discharge and no longer be chargeable.

---

[8] Your mileage may vary. All distance figures are highly dependent on surroundings.

- Get up high. The higher your antenna, the further the radio can see. Height extends the radio horizon.
- If you heard a signal well, but now it is weak or distorted, try moving. You may be in a radio shadow or receiving the signal on multiple paths as it bounces off objects. Moving just a few feet can make all the difference.
- The antennas shipped with HTs are usually shortened versions of what physics would require[9]. You can improve performance by substituting a full-sized antenna if the radio will allow it.
- GMRS and CB allow for directional antennas.
- Hold your antenna vertically to avoid cross-polarization.
- Elevate your antenna at least 15 feet. A painter's pole is good for this purpose.

The chief advantage of the unlicensed services is that there are many users already in place, and equipment is inexpensive. These radios are suitable for chatting around town to hear what others are experiencing. If you are going to set up a cross-town link, this may be all you need.

RECOMENDATIONS

A used mobile CB and magnetic mount antenna to stick on the car are inexpensive and provide a mobile communication post for local contacts.

A GMRS radio could be part of your kit as it runs higher power, can access repeaters, if they are available, and communicate with FRS radio users as well. GMRS will not be as crowded as CB.

---

[9] See the chapter on Antennas. A full-wavelength in meters is 300/frequency in MHz. Take one-quarter of that for the length of a full-size quarter-wave vertical in meters. 1 meter = 39.4 inches.

COMPARISON OF RADIO SERVICES					
Service	**FRS**	**GMRS**	**MUR**	**CB**	**Ham**
Channels	22 at 462 Mhz - 467 Mhz	22 FRS + 8 Repeater at 462 MHz to 467 MHz	55 at around 151 Mhz	40 at around 27 MHz	Unlimited within ham bands
Power	Channel 1-7 2W Channel 8-14 .5W	Channel 1-7 5W Channel 8-14 .5W Channel 15 - 22 50W	2W	4W AM 12W SSB	200 - 1500W
Antenna	Fixed	Separate allowed	Max height 60′	Separate allowed	Separate allowed
Line of sight	Yes	Yes but extended coverage with repeaters	Yes	Yes with sporadic ionospheric skip	VHF/UHF yes HF unlimited

LICENSED RADIO SERVICES

As a youngster, there was little entertainment. Three channels of black and white TV don't occupy you for long. I got a shortwave receiver one Christmas and started listening to international broadcast stations. Tuning across the bands, I heard people talking to each other. That was how I discovered Amateur Radio.

I knew immediately that was something I wanted to do. Listening to distant stations, fading in and out of the static, sometimes with distinctive polar flutter, but always with a sense of magic, planted a seed. If I became a Ham, I could graduate from a passive listener to a participant.

K4IA as WA4TUF in 1967

Nobody knows where the term "Ham Radio" came from, but you'll often hear it used to describe the Amateur Radio Service. I suspect people use "Ham" because they have a hard time spelling amateur. Just kidding.

There are over 800,000 Amateur Radio operators in the United States alone with over 30,000 new licensees every year. There are about 3 million Hams worldwide. They strive to be knowledgeable radio operators ready to bring their equipment and expertise to public service or emergency uses. All this expertise and equipment didn't cost the U.S. government a dime.

Amateur Radio advantages The Amateur Radio service has many advantages over unlicensed services. Here are some very important distinctions:

- A wide variety of operating frequencies so you can adjust for propagation.
- Privileges on the shortwave bands so you are not limited to line-of-sight communication.
- Maximum power of 1,500 watts.
- Much more sophisticated and capable equipment.
- Established operating protocols, message nets, and infrastructure.
- Amateurs are educated. You must study and pass a test for your license.
- Many UHF and VHF repeaters are available to extend the range of HTs.[10]

[10] See the chapter on Repeaters.

OBJECTIONS TO LICENSING

The major objections I hear are: "It costs too much to build a station," "I don't need a license to push a button and talk," and "It is too hard, and I can't learn Morse code." Let's deal with these.

It costs too much. You do not need to spend mega-thousands of dollars. Amateur Radio isn't cheap but, for the cost of a video gaming system or smartphone, you can get on the air. Amateur Radio costs nothing compared to SCUBA diving, golf, or a bass boat.

You probably have well over $1,000 invested in freeze-dried gruel you will never eat, equipment you will never use, and ammunition you will never shoot. Amateur Radio equipment won't sit idly under your bed and can provide years of enjoyment even if the worst never happens.

Take comfort in the fact Amateur gear holds value and consider the annual cost. If you choose to go all-out with new equipment and sell in five years, a $1,500 transceiver might fetch 75% of the new price. That would be a discount of $375 for five years of service or $75 a year. That amount won't buy much golfing.

There are active markets in used Ham gear on websites such as eBay, eHam.net and QRZ.com.

Used equipment depreciates even slower. Pay $500 for a used transceiver, and you might reasonably expect to sell it for very close to that in a few years. Say, you sold it for $450 after two years; your cost was only $25 a year.

We'll discuss specific equipment in another chapter.

I don't need a license. Anyone can buy a radio and mash a button to talk, but will you know what to do when the time comes? Can you communicate effectively? The path to being prepared is preparation. Getting a license will teach you the basics of how the equipment operates and what frequencies, antennas, and times are best to communicate.

Once you are licensed, time spent on the air is fun and good practice for whatever level of disaster you anticipate. Most radio clubs offer training and assist in public service events that serve as practice runs for emergency communication incidents.

Sure, in a real crisis, the authorities may have too much on their plate to bother with enforcing the Communications Act, but that is not the point. If you don't know how to operate the equipment, it is just a brick. You might as well wrap your message around that transceiver and throw it. That is how far you will get.

It is too hard, and I can't learn Morse code. Morse code is no longer a requirement for any class of Amateur license.

And, no, the tests are not too hard. I have taught many classes and taken people who have no math or science background through the material. They all pass the tests. Eight and eighty-year-olds have done it.

We'll talk about the tests in another chapter. You will want to get my EasyWayHamBooks license manuals.[11] Start with "Pass Your Amateur Radio Technician Class Test - The Easy Way."

---

[11] EasyWayHamBooks.com and Amazon

CLASSES OF AMATEUR RADIO LICENSES

The Federal Communications Commission regulates airwaves in the United States, and you must obtain a license from the FCC to operate Amateur Radio.

There are three amateur license classes: Technician, General, and Amateur Extra, each conferring more privileges. The privileges break down roughly as follows:

Technician

- UHF and VHF amateur bands above 50 MHz with some limited amateur HF privileges. There are proposals to expand the HF privileges.
- Technicians are limited to 200 watts output on HF.
- Mainly for local communications.

General

- All Technician plus most amateur HF frequencies.
- A portion of each HF band is reserved for Extras.
- 1,500 watts maximum output power.
- Local and worldwide coverage possible

Amateur Extra

- All frequencies allocated to the amateur service
- 1,500 watts maximum output power.

AMATEUR RADIO LICENSE TESTING

How hard is the Technician test? It is not hard. There is no heavy-duty algebra, calculus, or trigonometry. Don't let the material intimidate you. There is no Morse code required for any class of Amateur license. Morse code is still very much alive and well on the amateur bands, and there are many reasons you might want to learn it in the future. Don't worry about it for now. The General test is a little tougher but not much.

The Technician and General tests are each 35 questions from a pool. The Technician pool has 424 possible questions, and the General has 462. The questions and the answers on your test are word-for-word out of the published question pool. You know what will be on the test and can study with confidence you are focusing on the right material.

The pool is large, but only about one in a dozen questions will show up on your exam, one from each of the test subjects. Many questions ask for the same information in a slightly different format, so there aren't 424 or 462 unique questions. You can miss 9 of the 35 and still pass.

Do you know what they call the medical student who graduates last in his class? "Doctor" – the same as the guy who graduated first in the class. My point is: all you need is to pass, and no one will know the difference. To pass, answer 26 out of 35 correctly.

The questions are multiple-choice, so you don't have to know the answer, you just have to recognize it. That is a tremendous advantage for the test-taker.

The best way to study for a multiple-choice exam is to concentrate on the correct answers. Find a word or phrase you can remember. When you take the test, the correct answer should jump right out at you. I take that approach in my EasyWayHamBooks license series.

If you don't know the answer, eliminate the obviously wrong answers. Then make an educated guess. There is no penalty for guessing wrong, and you greatly increase your odds by eliminating bogus responses.

You can find classes listed at http://www.arrl.org/find-an-amateur-radio-license-class or check with your local Amateur Radio club. You are not required to attend a class, but may find it helpful.

Volunteer Examiners organized under the authority of the FCC, administer Amateur Radio testing. VEs are local hams who devote time to help you get your license. In most areas, you can find a VE group testing at least once a month. There are 14 accredited VE organizations, but ARRL and W5YI are the biggest. Look on the licensing tab of ARRL.org or W5YI.org to find a test site.

On test day, you should bring a picture ID, your Social Security Number (not the card, just the number) or your FRN number,[12] a pen and pencil, and about $15. You need to ask ahead to get the exact amount and see if they prefer cash or check. The modest exam fee reimburses the VE's costs — no one profits except you. If you bring a calculator, you must clear the memories. Turn off your cell phone and don't look at it during the exam.

---

[12] Many VEs prefer you use a Federal Registration Number (FRN) from the FCC instead of your Social Security Number. The FRN is used only for your communications with the FCC. Search "FRN FCC" to get the link to the FCC website where you can register. Then, bring your FRN number to the test session.

You will complete a simple application (Form 605) and might save a few minutes by downloading the form from the Federal Communications Commission (FCC) website and filling it out ahead of time.

Your Volunteer Examiner (VE) team will select a test booklet for you. Ask them for the easy one – that's good for a cheap laugh. They have no idea what questions are in your booklet. You also get a multiple-choice, fill-in-the-circle answer sheet. The answer sheet has room for 50 questions, but you stop at 35 for Technician and General. You can take notes and calculate on the back of the answer sheet – not in the booklet. VEs re-use the test books. There is no time limit, but most finish in 30 minutes or less.

The VE team will grade your exam while you wait and give you a pass/fail result. Don't ask them to go over the questions or tell you what you missed. They don't know and don't have time to look.

When you pass, there is a bit more paperwork, and you walk out with a CSCE – a Certificate of Successful Completion of Examination. The team processes your Form 605 and CSCE, and your personal call sign appears in the FCC (Federal Communications Commission) database in a few days. Once you see it there, you are good to operate. Congratulations.

You can print a copy of your license from the FCC website. The license is good for ten years, and you can renew without testing again.

HOW YOU SHOULD GET STARTED IN AMATEUR RADIO

Here are a couple of tips to get started in Amateur Radio:

- Join your local Amateur Radio club.
- Find an Elmer.
- Attend Hamfests.
- Participate in Field Day.
- Get on the air.

Join Your Local Amateur Radio Club You can find lists of local radio organizations at HamDepot.com or ARRL.org/find-a-club. There are clubs everywhere and most host a repeater. Listen to the repeater to learn more about the group. Attend a meeting. Clubs are anxious to receive new members and encourage new Hams. Don't be intimidated if the discussion is over your head; you just need time to catch up.

Go and meet some folks. Participate on the repeater and some club activities. See if the group is a fit for you. Some emphasize public service, providing communications for marathons, bike races, and other events. Other clubs may be interested in contesting or just eating breakfast. If you don't feel the love, try another organization.

Find An Elmer The local club will allow you to rub elbows with experienced HF operators and learn from them. Joining a local club is a great way to fulfill the second piece of advice: find an Elmer. An Elmer is a mentor to coach you. This isn't a formal commitment. You're not "going steady" or anything like that. Look for someone who seems friendly, more knowledgeable, and willing to answer questions. Your Elmer is your "go-to" person when you are stumped or don't understand the jargon.

You might find more than one Elmer. One is the guru of antennas; another is a contesting fanatic, and another understands computers. You will find most Hams willing to share their expertise, so don't be afraid to ask.

Attend Hamfests A Hamfest is a Ham Radio oriented gathering to swap/buy/sell, attend forums, meet for an eyeball QSO[13], demo equipment, and have a good time. It is a convention that mixes a flea market, museum, vendors, and camaraderie. You will often find reasonably priced used equipment along with connectors and cables that aren't economical to ship if ordered online.

Hamfests can be elaborate multi-day affairs attended by tens of thousands such as the annual Dayton Hamvention in May, the Visalia, California DX Convention in April, or the Orlando Hamcation in February. There are also numerous local Hamfests and low-key tailgate "junk-in-the-trunk" gatherings. You can find lists of upcoming events at arrl.org/hamfests-and-conventions-calendar.

Participate In Field Day Field Day is a long-standing Amateur Radio tradition. The last full weekend in June, Hams take to the great outdoors and operate using emergency power.[14] The goal is to set up a station and contact as many other stations as possible during the 24-hour activity.

This event is a combination of emergency-preparedness training, contest, picnic, and campout. Your local club probably participates in Field Day.

---

[13] Meet face-to-face.

[14] There are other operating categories as well but this is the main one.

HOW TO GET STARTED

Field Day is an opportunity for you to see antennas erected, stations set up and operators operating. You can take part as a helper or operator.

The rules allow for a Get-On-The-Air (GOTA) station restricted to new or "generally inactive" Hams. An unlicensed person may operate under supervision. Ask about the GOTA station and take advantage of the opportunity. Don't miss Field Day. It is an exciting time.

Get On The Air The last piece of advice is, "Get on the air." The way to get started is to start. No matter how modest your station, you can make contacts. Operating tips and techniques are coming up in another chapter.

If you don't have a station, ask a local Ham to let you visit his shack. Most Hams are proud to show off their equipment introduce you to the airwaves.

ASSEMBLING A HAM, FRS OR GMRS VHF/UHF STATION

EQUIPMENT FOR A VHF/UHF STATION

Transceivers are the rule for VHF/UHF equipment. That is, the transmitter and receiver are in one package. The transceiver can be an HT, base (desktop), or mobile (car-mounted) station.

The Handie Talkie (HT)

HTs are the basic VHF/UHF station. HTs generally output 5 - 10 watts and get power from rechargeable batteries. The 10-watt version needs more battery and will be heavier and bulkier. There is very little difference on the receiving end, so the trade-off may favor a 5-watt model.

Some HT manufactures provide auxiliary battery packs that accept standard AA batteries. Those are handy as a quick replacement for a rechargeable while it is recharging. Maximize battery life by turning down the transmit power to the minimum needed to make contact. Minimum power is an FCC rule.

Every HT should have three power sources: internal battery, backup battery, and wired. The wired connection is usually a cigarette-lighter plug.

The modulation mode is FM. FRS and GMRS HTs operate on a single band. Some ham HTs operate on a single band, 144 MHz, but most are dual-band adding 440 MHz. The dual-band versions don't cost much more, so get a dual-band.

You can program repeater[15] frequencies, offsets, and tones through the HT keypad, but it is much easier to

[15] See the later chapter on Repeaters

use a computer program. Find your local repeater information with a smartphone APP called Repeaterbook or at RepeaterBook.com. Use the computer program that came with your radio or look for the free program "Chirp."

The accessories you should order include:

- Programming cable to connect to a computer and set up the radio.
- Full-size outside antenna.[16]
- Full-size attachable antenna
- Extra rechargeable battery and/or AA battery carrier.
- Earphone. The little speakers can be hard to hear in a noisy environment.

Base/Mobile Station

An HT alone is not enough radio. It may be handy for walking around, but you need something more powerful for effective communications. That requires a Base/Mobile station. VHF/UHF gear tends to be small and fits on a desk or mounted in a car. There is another chapter on mobile operation with more detail.

The typical Base/Mobile rig will output 15 – 50 watts for GMRS and up to 100 watts for Ham.

Some Ham models can be a cross-band repeater. Here is how that might work. The 440 MHz band works better than 144 MHz for penetrating walls. When our local club supports the Special Olympics, we are inside a school building. An HT on 440 MHz communicates to a cross band repeater in the parking lot, and that relays to the big repeater operating on 144 MHz.

---

[16] Your car is a metal box and if you try to operate an HT from inside the car, your signal will be very weak. Get an outside antenna. One type clips over your window.

CHOOSING THE EQUIPMENT

The big-four manufacturers are Icom, Kenwood, Yaesu, and Alinco. Motorola produces expensive commercial-grade gear.

There is a gaggle of inexpensive Chinese radios with names like Wouxun, Baofeng, TYT, and Hesenate. They all look alike and probably come from the same factory. China makes almost all of these radios, including the big-name brands. They all have similar functionality. The differences are in spectral purity, ruggedness, and overall quality.

You can order through Amazon or the retailers listed in the chapter on Assembling an HF station.

I am not going to recommend particular models or manufacturers. Manufacturers introduce new gear every year, and it would be too hard to keep up. We all have our favorites, and some people like Fords while others favor Chevys. As my beloved Nana used to say, "That's why they don't make all ice-cream vanilla." Here are some of the criteria you might apply when making a choice of equipment for your VHF/UHF Ham or GMRS station.

HOW MUCH DO YOU WANT TO SPEND?

UHF/VHF gear is inexpensive compared to HF. This is largely because it is lower power and limited in frequencies and modes. That is not to say it can't get expensive.

There are new digital modes that add considerable cost to a basic radio. They go by names like D-Star, DMR, and System Fusion. No one system is supreme. Remember the battle between Sony BetaMax and VHS tapes? Do some investigating before you buy. Check with Repeaterbook.com to see what is supported in your area. A simple FM radio is all you need to start.

VHF/UHF STATION

The majority of Amateur Radio activity is on two-meter FM. Check with the locals before you spend money on bands and features you will never use.

Used Equipment

Basic used VHF/UHF equipment is inexpensive. By "basic," I mean one or two bands, FM, either an HT or a small mobile rig. A club member might be willing to give you something or let you borrow it.

Things to watch out for when buying used:

- HT battery is probably dead and can be expensive to replace.
- Missing manual.
- Broken antenna.
- Missing parts (microphone, mobile mounting bracket, charger, or power cable.

Look on eBay or Amazon for replacement parts. Manuals are usually on the manufacturer's website.

New Equipment

Everyone needs at least one HT. Cheap Chinese versions start at less than $50 but may be of questionable design and offer poor performance. You will have to navigate through a manual written in "Chinglish" and figure out how to program the frequencies. It is easier using a computer and programming cable. If the manufacturer doesn't provide an interface, download a free program called "Chirp."

A dual-band base/mobile FM radio is less than $200. Add $50 for an antenna to stick on your car. If you want to use it at home, you will need an outside antenna ($50 - $100) power supply ($100) (and cable to feed it ($75). These are very rough estimates, as every installation is different.

AMATEUR RADIO HF STATION EQUIPMENT

An Amateur HF station is usually a transceiver, a transmitter and receiver in one box.

TRANSMITTER

When you select a transceiver, the transmitter section is not nearly as important as the receiver, and the receiver will dictate your choice.

Fortunately, this is not much of a dilemma. Most transceivers deliver 100 watts, and 100 watts are the same no matter where they originate. You can argue about phase noise, clean signals, and key clicks but modern equipment has (mostly) slain those dragons.

The vast majority of Hams operate HF with 100 watts and 50 watts on VHF or UHF. In most instances, that is sufficient power and will provide communications over reasonable paths. Amplifiers can take you up to 1,500 watts output power.

Bells and whistles drive up the price. If you are on a budget, just accept whatever the transmitter portion has to offer. The basic radios will suffice. Concentrate on buying the best receiver.

Helpful but non-critical, features include:

- Built-in antenna tuner.
- Built-in power supply (rare with modern equipment).
- Additional band coverage (6 meters and VHF/UHF).
- Built-in speech processor so you can tailor your audio.

- Built-in voice and CW keyer to play back pre-recorded messages, although you can do this with outboard equipment.

RECEIVER

When comparing radios, you should concentrate your attention on the receiver. That is the most critical part of any transceiver.

Sensitivity is not the issue as they are all about the same and will receive at the natural background noise floor. What sets an excellent receiver apart from a mediocre one is selectivity and particularly selectivity to resist intermodulation by-products (IMD) from nearby strong signals. IMD is an unwanted mixing of signals that sound like growling or distortion.

The latest tool to defeat such interference is called a "roofing filter." The roofing filter is very early in the receiver chain to reduce the breadth of signals reaching the later, more sensitive circuits. A narrow roofing filter will also reduce AGC[17] pumping caused by loud close-by signals.

If you are not operating under crowded band conditions, you may never notice the need for a roofing filter. But, now you know what they are.

Roofing filters or not, you need filtering down the chain. Crystal filters are one answer. Look for a receiver that accepts filters appropriate for the mode. Some low-end transceivers ship with one moderately wide SSB filter, around 2.7 kHz, and don't have a way to add additional filters. Even if you don't order the narrower filters right away, it is nice to know you can add them later.

---

[17] Automatic Gain Control operates to level out the audio. AGC will automatically reduce the gain on a loud station to protect your ears.

Consider a narrow 1.8 kHz filter for SSB and a 500 Hz CW filter. New, these cost about $150 each, so a used radio that is "fully filtered" may be a bargain.

These are crystal filters. Another type of filter is Digital Signal Processing. Digital Signal Processing converts the signal to numbers (1's and 0's), looks for patterns, and then manipulates the numbers to cancel out interfering signals and noise. The math is staggering, but microprocessors in the radio make it possible.

A crystal filter has a fixed width. If the filter is 1.8 kHz wide, that is all it will do. DSP can be molded to fit the need and is continuously adjustable, so you are not limited to preset crystal-filter bandwidths. DSP processing was expensive, but the price is coming down, and you will find it in modern moderate-level transceivers. Some older receivers have audio DSP. Newer units have DSP in both the audio and RF stages for exceptional filtering and noise reduction.

OTHER ACCESSORIES

The modern Ham shack includes much more than a transceiver and antenna. There are dozens of available accessories. You do not need all these gadgets.

Antenna Analyzer Measures how well your antenna system matches your radio. An analyzer is different from an SWR[18] meter because the analyzer generates its own low-power signal and can display a range of measurements over different frequencies. The SWR meter uses the transmitter's power, only reads one

---

[18] Standing Wave Ratio is a way of expressing how well the antenna system matches the radio. Ideal is 1:1 but anything less than 2:1 is acceptable and even 3:1 is usable.

frequency at a time, and only measures forward and reflected power, deriving from that the SWR.

Antenna Switch Handy for changing antennas and a lot easier than screwing and unscrewing coax connectors. Many provide a center "ground" position, but you should not rely on that for lightning protection. See the later section on "Grounds and Lightning Protection."

Anderson Power Poles These are a standard plug-in power connector for attaching a power supply and equipment.

Baluns and line isolators If you experience RFI (Radio Frequency Interference), these may be the cure. See the Chapter on "Grounds and Lightning Protection."

DC Power Distribution Panel You might have several pieces of equipment to hook into your DC power supply. A distribution panel with multiple connectors makes it easier to connect them all. Some use Anderson Power Poles for convenience.

Dummy Load A large resistor, sometimes air-cooled and sometimes in a bucket of oil. You test or tune-up into the resistor, so your signal does not go out over the air.

Filters There are many different kinds of filters serving different purposes. AC line filters reduce voltage spikes and block noise on the AC line. Audio filters can use tuned circuits or digital signal processing to reduce noise or narrow the audio passband to reduce adjacent interference.

RF filters suppress RF energy outside their design frequency. If you have a powerful AM radio station nearby, an RF filter might help avoid it overloading your receiver.

Headphones You don't need to spend a fortune on high-fidelity studio equipment because deep bass and tinkling highs don't matter. The most important specification for headphones is that they must be comfortable and not crush your head in a vise.

There are three basic headphone designs, and you might want to own all three for different occasions. Over-the-ear models cover the ear completely and block out external noise. They can become uncomfortable if worn for too long but are helpful in a noisy environment.

On-the-ear designs sit on top of the ear and don't provide much in the way of noise-blocking but are more comfortable. I prefer on-the-ear unless I am in a noisy environment.

Earbuds fit in the ear and do a better job of noise blocking than on-the-ear designs and are more comfortable than over-the-ear. I like earbuds for long sessions.

Multi-Meter These handy devices measure volts, amps, and resistance. You should have at least one. The cheap versions (under $8) are not laboratory-grade but are sufficient for most of our needs. Harbor Freight often offers them free to get you into the store. I pick one up whenever I can.

Speaker The speaker in your transceiver is probably not ideal. It fires upwards instead of at you. It is also a bit small. Wide fidelity is not important, and you might find that built-in speaker sufficient. An external speaker will improve your reception. You don't need to pay a premium price to get a speaker that matches your rig. When they say it matches, they only mean it is painted the same color.

Speech Processor A speech processor lets you tailor the audio to increase readability by adjusting the frequency response to emphasize the mid and mid-high frequencies that carry the most information.

SWR Meter The SWR meter goes between your transmitter and the antenna to measure forward and reflected power as you transmit.

Voice Keyer A voice keyer allows you to record a message and play it back with the push of a button. You set it to auto-repeat, mash the button, sit back, and wait for an answer.

The best accessory is an assistant as shown on this QSL card from the Canary Islands. Hams exchange postcards, called QSL cards, as written proof of their contacts.

ASSEMBLING AN AMATEUR RADIO HF STATION

The HF station is a bit more complex than the typical VHF/UHF station. You have more frequencies and modes, and the antennas are much larger.

HOW MUCH DO YOU WANT TO SPEND?

Moderate Used A moderate level used station would be an all-mode (AM, SSB, CW, Digitial) transceiver five to fifteen years old. It will cover the basic HF ham bands and may receive in the shortwave bands (called "general coverage"). This equipment is modern enough to give plenty of service but won't be on the cutting edge of technology. It would not have been top-of-the-line even in its day. The higher end of the price range would give you something newer or a more fully equipped older rig. You can expect computer control, good filtering, solid-state, and an antenna tuner but not the latest-and-greatest of any of these features.

Moderate Used Station

Used HF Transceiver	$300- $1,000
Used Antenna Tuner	$100
Multi-Band Wire Antenna	$75 - $150
Feedline	$50 - $100
Used Power Supply	$100
Misc	$100
TOTAL	$750 - $1,550

Specific Considerations
There have been hundreds of different radios produced in the last 15 years. I can't possibly list a fraction of the candidates.

- Check the reviews on eHam.net.
- Look for features like a built-in antenna tuner, and a general-coverage receiver.

- Ask the seller about included accessories (microphone, cables, and manuals).
- Has the radio been in a smoking environment?

Moderate New Station

If you decide to go with all new equipment, the price escalates. The following chart describes an all-new station with moderate level components. This isn't the most expensive gear available. It is adequate, especially for a beginner. After you have some experience and decide on your style of operating, you may want to upgrade, but many Hams stay in this category for years.

Moderate New Station:

Mid-range HF Transceiver	$500- $1,500
New Antenna Tuner[19]	$150 - $250
Multi-Band Wire Antenna	$75 - $150
Feedline	$50 - $100
Power Supply	$125
Misc	$100
TOTAL	$1,000 - $2,225

As you can see, there are lots of options. Anything you can beg or borrow will reduce the initial cost. These figures are very rough estimates designed to get you in the ballpark.

Specific considerations

The bottom of the price range will sport a general-coverage receiver, 100 watts, all-mode, HF only (160 meters – 10 meters), and no internal antenna tuner. As you move up, you will add the antenna tuner and more advanced features like digital signal processing and filtering.

---

[19] Maybe, you won't need one if your transceiver has an internal tuner. See more in the chapter on antennas.

Sellers and Dealers
There are several highly reputable and very helpful dealers online. You can visit their retail stores to spin the knobs on new gear. Most offer free shipping on orders over $100, and they will be glad to send you a catalog.

Equipment dealers include:

- Ham Radio Outlet www.HamRadio.com with 13 brick-and-mortar stores across the country.
- DX Engineering www.DxEngineering.com in Tallmadge, OH
- Gigaparts www.Gigaparts.com has stores in Huntsville, AL and Las Vegas, NV
- Universal Radio www.Universal-Radio.com is in Worthington, OH
- Ham City www.HamCity.com is in Gardena, CA

USED VS NEW

Much like computers, radio equipment has undergone incredible technological advances. However, radio itself hasn't changed, and there is a lot of 50 to 60-year-old gear still in service. You can save some money with used equipment, but what are the considerations when buying used?

Vacuum tube gear is a collector's item. Some of it still works just fine. I have a 50-year old Drake transceiver I enjoy very much. However, parts and tubes can be hard to find, and tube radios are projects, much like an antique car. I would not recommend you start with vacuum tube gear unless someone gives it to you.

Finding replacement parts for old gear can also be a problem. By "old" I will arbitrarily say, 20 years. It's not that you can't find a particular resistor; it is the no-longer-made band switches, displays, and mechanical pieces that are "unobtainium." Older radios lose alignment, and parts may change their value with time. Keeping up old equipment is a hobby unto itself, and unless you are comfortable troubleshooting and fabricating, you might better stay away.

Having warned you off older gear[20], I must add a qualifier. There are many vintage transceivers still performing excellent service. Take one as a loaner or a gift but don't pay a lot, $250 - $350 maximum. Make sure it is working properly because the cost of a professional repair (with shipping back and forth) might be more than the radio cost.

Ten to fifteen-year-old gear is priced temptingly at about half its new cost. Future depreciation will be slow. It will sell for close to what you paid as long as nothing breaks. Check the reviews online (eHam.net). See if a particular issue plagues the radio such as failing displays, bad final transistors or unobtainable replacement parts.

Five-year-old radios are practically new and sell for 75-80% of their original price. These can be a real value as long as the radio works properly and doesn't stink from being in a smoker's shack.[21]

There was a time when you could trust a Ham not to sell a problem radio without full disclosure. I don't know if you can count on that courtesy any longer. Know your seller. Buying from a local Ham is probably safe. Don't expect a guarantee. If you want a warranty, buy something new.

Buying used radios online is riskier though I have experienced good luck with sellers who had high positive feedback. PayPal and eBay have strong buyer

---

[20] Sometimes called "boat anchors."

[21] The "shack" is a radio room.

protection policies, so you do have recourse if the radio is not as represented, dead or damaged on arrival. They can't help you if the gear blows up after a week, even if you suspect the seller knew something was going bad.

Other online sites don't have the same buyer protections, so understand the terms before you buy. Be careful of online scams. If the price is too good to be true, it is.

Try to borrow gear. The local club may loan equipment to new Hams. A member might have a radio sitting in the bottom of his closet, and be willing to let you use it. Finding equipment is another good reason to join your local club.

If you do borrow, be a good steward and treat the gear carefully. Have someone show you how to use the radio and spend time in the instruction manual before you do any damage. The manual is often available online. Get a clear understanding of liability. Is the responsibility, "If it breaks in your possession, you bought it" or "I understand it is old and feeble so whatever happens, happens?"

Agree, in advance, on the terms of your loan. I once let out a radio with no clear understanding, and it took me three years to get it back. If I loan again, I will take a picture of the borrower with the equipment and a sign that says, "Return by Jan 1, 2019." That gets the point across and helps everyone remember. On January 1, I can send him the picture as a hint. I am only half kidding.

TRANSCEIVER FRONT PANEL CONTROLS

The array of controls on a modern transceiver is overwhelming. Large panels allow room for lots of knobs and buttons for your adjustments. Small transceivers have very few controls and rely on multiple button pushes or menus to dig down into the parameters you might adjust. Here is a list of standard controls and what they do. I use the terms "switch" and "button." You flip a switch, and you push a button, but the function is the same.

AF Gain A volume control for the audio amplifier. Compare to the RF Gain control for the radio-frequency stage amplification.

Antenna Selector Your transceiver may come with more than one antenna jack on the back. You can select the antenna from this switch or button.

AGC (Automatic Gain Control) Suppose you are listening for a weak signal and have the gain turned up. Then you tune across a very loud station. AGC protects your ears by leveling out the volume.

AGC usually has two settings, fast and slow. "Fast" will recover quickly while "slow" will hold down the amplification longer. You use different settings depending on the conditions. Voice communications usually sound better with "slow."

Attenuation Attenuation reduces all energy coming into the receiver, both noise, and signals. It is very useful for the lower bands where a high noise level can overwhelm the sensitive circuits.

Antenna Tuner Properly called a conjugate matching device; the antenna tuner adjusts the load to match the transceiver. You will hear the clattering of relays as an automatic tuner works to find the right solution. Don't worry; it is not broken.

Band Switch Rather than spin the frequency knob to go from 40 meters to 20 meters, one push changes the band, usually returning to the last frequency you used on the new band.

Compression Your voice has loud and soft inflections. Some words are naturally louder than others. Compression works to fill out the audio by boosting the less-strong words and syllables. You adjust the amount of compression to get that boost, but not so much as to add distortion.

CW Speed Most transceivers have a built-in keyer. This control adjusts the speed of the Morse Code sent from paddles.

CW Pitch The pitch is the musical tone. If you have it set to 650Hz, when you hear the received audio signal at 650 Hz, you will be on the same radio frequency as the sender or "zero-beat." Pick a tone you like to hear and teach your ear to remember it.

Frequency Entry Enter the frequency on a keypad rather than spin a dial. Usually, you have to push an additional button to activate the change. It may be called "Enter."

Filter Width May be fixed or continuously variable. Narrower filters restrict the range of signals heard.

Filter Shift (PBT, passband tuning) Alters the shape of the filter to favor frequencies above or below the center of the filter.

Mic Gain Adjusts the volume of the microphone input to the transmitter. Talk close to the microphone and adjust the mic gain per your transceiver's instructions.

FRONT PANEL CONTROLS

Most Hams "close talk" the microphone within an inch of their mouth. If you speak too far away and turn up the mic gain to compensate, you will also pick up background noises such as barking dogs or fans.

Message Playback Some transceivers have a built-in recorder for voice and CQ play-back. You record a message and play it back with one button push. A recorded message makes it a lot easier to call CQ and helps you get over mic fright. Pushing a button is less intimidating than calling.

Memories Just like your HT, you can program frequencies and modes. Programming makes it easier to return to a particular frequency and is handy if you check into a net regularly. Set a memory slot for the top, middle, and bottom of a band to move around without spinning the dial.

Mode Selects AM, USB, LSB, CW, and Digital modes.

Monitor You can listen to your transmitted signal through this function. I like to keep the monitor volume low, so my ears stay sensitive to receive.

Noise Blanker Activates a circuit designed to mute the sound of repetitive noise such as a spark plug or electric fence.

Noise Reduction Filters out random noise. There may be multiple settings to deal with differing conditions. Very aggressive noise reduction can introduce distortion, so this adjustment is by trial and error.

Notch Reduces or "notches" out an offending signal. Auto-notch will seek out and silence a carrier[22] near

---

[22] If you are listening to SSB and someone decides to tune up near your frequency, you will hear the tone of his carrier. Notch will cut it out.

your frequency. You adjust a manual notch to do the same thing.

Power The on/off button, of course, but there is also a control to set how much power you transmit.

Pre-Amp Turn on an additional amplifier applied to signals before they enter the receiver. Use this sparingly and only on higher frequencies with low background noise, or you will overwhelm the receiver.

QSK When activated, your transceiver will switch from transmitting to receive almost instantaneously. QSK is very helpful on CW as you can hear the receiver between sending your characters.

Receiver Incremental Tuning (RIT or Clarifier) Tunes the receiver without changing your transmit frequency. If someone called you slightly off frequency, you would use this to zero them in and "clarify" the signal.

If you adjusted the VFO to make the incoming signal sound better, you would change your transmit frequency and sound weird on his end. Then he would adjust, and the two of you would chase each other all over the band. Transmit in one place and use the RIT to correct for the off-frequency signal.

Reverse To change from USB to LSB or to change the side on which you hear a CW signal. On CW, this is an effective interference fighting too. Reversing the side will make the interfering signal further away from your tuned frequency. On SSB, stick with the convention for the particular band. It does no good to listen to an upper sideband transmission on a lower sideband setting. You won't be able to understand it.

Another use of the term "reverse" is to exchange VFO A and VFO B frequencies. That is a different control. (See below).

RF Gain Adjusts the gain in the radio-frequency section of the receiver. Turn the RF Gain down so you can just hear the background noise. This setting preserves the maximum dynamic range of your receiver. If a signal is below the noise, turning up the noise won't make it any easier to copy. Turning down noise is also less fatiguing.

Spot Injects the audio tone you selected with your CW pitch control to help you zero-beat the other station.

Squelch Silences your receiver when there are no signals, the same as on your HT. Squelch is not usually used on HF because it only triggers with signals substantially above the noise level. HF signals may be very close to the noise level and won't be loud enough to open the squelch.

Transmit Incremental Tuning Changes your transmit frequency while leaving the receive frequency the same.

Tuning step Determines how quickly the frequency changes as you spin the dial.

VFO Variable Frequency Oscillator is your tuning control.

VFO A / VFO B If your transceiver allows, you can have two frequencies active. You only transmit on one, but in split operation, you listen to the other. The **A/B** button would switch the two.

VOX Engages the voice operated relay, so you don't have to use a transmit/receive switch like the push-to-talk commonly found on a microphone. VOX facilitates hands-free operating.

VOX Delay Sets how long the radio stays in transmit after you stop talking. It can be very distracting if the delay is set too short.

VOX Anti-Vox Noise from your receiver speaker can fool the VOX into thinking you are talking and want to transmit. The AntiVox control is negative feedback that cancels out the received signal, so it doesn't trip the VOX into transmit mode.

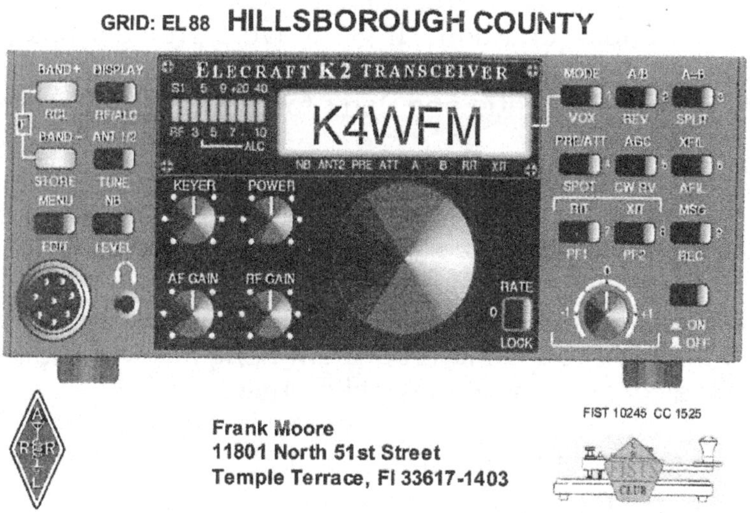

This QSL card shows the front panel of an Elecraft K2 transceiver. It features multi-use buttons. Push activates one function. Push-and-hold activates a second.

TRANSCEIVER REAR PANEL

The rear of a transceiver can be as confusing as the front. There may be a few controls hidden on the back for functions that don't often change such as microphone compression or VOX[23] delay. If you're looking for a particular control and can't find it, check the back of the radio. Also, look to see if it is under a trap door on top of the chassis.

The rear contains the connections, the goes-intas, and goes-outtas. There could be discrete jacks and cables for various functions, or they might be bundled in a DIN connector. A DIN connector is small and round with 5-8 pins. The plug is keyed so it will only fit the jack in one orientation. If you have a problem pushing it in, you probably don't have the key tab lined up correctly. Be gentle and don't force it.

You might not be able to see the rear of your equipment, or the only way to see it is to lean over the front and then everything is upside-down. It helps to take a picture or sketch and have it handy when you need to make changes. I put labels upside-down on the rear so I could read them hanging over the top. Those connectors all look alike otherwise.

Here's a list of what you might find on the back of a transceiver:

Accessory Jack Control other equipment or pass audio. The various pins may allow for additional audio inputs and outputs and packet or RTTY (teletype) signals.

ALC Automatic Level Control interfaces with an amplifier to assure the transceiver doesn't apply too

---

[23] VOX is voice operated relay. It keys the transmitter when you start to talk eliminating the need for a push-to-talk switch.

much power and overdrive the amp. If the amplifier senses it is over-driven, it changes the ALC voltage to reduce power from the transmitter.

Amplifier Keyer (may be called "key out"). A cable connected to an external amplifier to switch it to transmit mode when the transceiver is sending.

Antennas Connections for the HF and VHF/UHF antennas as applicable.

Fuse or circuit breaker In addition to a fuse in your power cord, many transceivers have a safety fuse or breaker. Keep a few spare fuses in your toolkit.

Grounding lug To attach your station grounds. See the chapter on "Grounds and Lightning Protection."

Key or paddles Input for CW operation. A key uses a mono plug for its two wires. Paddles use a stereo plug for ground, dits, and dahs.

Line in and Line Out Audio for digital modes and voice keyers or recorders. These may be a DIN plug or individual audio plugs.

PTT Separate push-to-talk switch. You may have a switch on your microphone that activates through the microphone jack. A separate PTT connector is for a footswitch or other hand switch.

Serial/USB port so the radio and computer can communicate.

Speaker output Plug in an outboard speaker.

SHACK DESIGN

If you decide to set up a radio room (shack), make it as comfortable and inviting as you can. I shudder to see Hams relegated to windowless, damp and dingy basements crouched among the cobwebs, behind the furnace, out of contact with the rest of the household. Radio is your pride and joy so come out of the cellar. Spouses don't hate your hobby. They hate your isolation. Be a part of the family.

I know you can't make a lot of noise in the living area. Isn't there a room somewhere on the main level where you can get away from the TV and have a little solitude? Do you have a den or sun porch? I like to look out the window and see the birdfeeder or the grandkids playing.

No matter where you are, brighten it up with adequate lighting and put a rug on the floor. Get a comfortable desk with enough room to spread out. I used a door on sawhorses as a kid. Now, office furniture is available in comparatively cheap, attractive and utile packages. Get a decent office chair and stop sitting on that old broken and unpadded kitchen stool. Make yourself comfortable so you can be a part of the family and maximize your radio time without killing your attitude, your posture, or your marriage.

The computer has become an essential element in the modern Ham shack, but we are assuming there is no power and no internet, so my picture on the next page is for demonstration purposes only.

Things that need adjusting should be within easy reach. I am right-handed so my transceiver is in the where I can twiddle the knobs with my right hand. The radio belongs on the desk level so you can rest your arm while tuning. Having the radio above the desk is very tiring and uncomfortable. My Morse code key is also to the right.

I have two small speakers on either side of the transceiver. They provide excellent stereo sound. When the going gets tough, I pull out the cans (headphones).

Computer monitors dominate in my shack.

As a general rule, anything that does not need adjusting can be out of reach. Bookshelves hold auxiliary equipment such as the antenna tuner, wattmeter, and digital interface; all of which are within reasonable but not immediate reach. I don't need to adjust those components often, but I do need to see the displays.

The bookshelves were originally too tall, so I cut six inches off the bottom and rearranged the shelves.

The power supply is on the floor where it belongs. I use my big toe to turn it off and on, but most of the time, I leave it on. If you put the power supply next to your transceiver, it will induce hum and other noise.

It also takes up room at your operating position. You never adjust the power supply, so why should it be on the desk?

An L-shaped desk makes two monitors easier to place as seen in this Irish station.

EMP

An EMP or electromagnetic pulse is a scary-sounding menace. A sudden release of electromagnetic energy induces a current in wires and components. If the current is strong enough, it can damage the equipment. The sun, lightning, or a nuclear explosion could generate such a pulse, and the fear is that an EMP will destroy all electronics and collapse the power grid.

The most famous electromagnetic pulse, known as the Carrington Event, occurred in 1859 when a huge solar storm sent a massive electronic energy burst to Earth. The northern lights, normally limited to areas around the poles, were visible almost to the equator. Telegraph operators reported sparks across their keys and electric shocks. In 1989, Quebec suffered an electrical system failure leading to a nine-hour blackout. I experienced an EMP from a nearby lightning strike that destroyed my receiver.

Disconnect The common denominator to the three examples was a long piece of wire: an antenna, telegraph, or power line. The energy from the pulse transferred to the line. Step number one to avoid EMP damage is to disconnect your equipment when not in use. A damaging pulse can't come down the antenna wire or through the house mains and damage your radio unless they are connected. Disconnecting is a good habit to cultivate.

Shield Step number two is to shield your equipment. The radio's metal case is a shield. Connect the case to ground as described in the next chapter on grounding. If that is not practical, say with an HT, you can store the equipment in a small metal trashcan and ground the can.

EMP

Predicting an EMP is difficult, but we do get notice of solar and lightning-related EMPS. When there is a solar event, the first manifestation is a solar flare. X-rays from the flare travel to Earth in 8 ½ minutes, the speed of light. This will disrupt the ionosphere and shut down long-distance radio communications. Usually, the effect only lasts a day or two. Scientists can predict what is to come based on the intensity of the X-Rays.

A coronal mass ejection accompanies the flare. The CME is a far more damaging burst of plasma, hot magnetically charged gas, and particles that take several days to reach Earth. If Earth happens to be in the path of the material, we get auroras. If severe enough, it can also damage satellites and disable electrical grids.

You can keep an eye on solar activity by monitoring SpaceWeather.com (assuming you have internet).

NOAA weather broadcasts, and the sound of thunder warn us of lightning. If you can hear thunder, you are within the possible lightning zone — time to shut down and disconnect.

GROUNDS, LIGHTNING AND EMP PROTECTION

A typical station has three "grounds." An electrical safety ground, a lightning ground, and a radio frequency (RF) ground. They fulfill three different functions but operate in harmony and must be interconnected. The fundamental concept is to avoid a difference in potential between components, and that means minimal resistance to both AC and DC.

ELECTRICAL SAFETY GROUND

The electrical safety ground is the third hole in an electrical outlet and connects the green wire back to the electrical panel ground. The plug wire connects to the case of the equipment, whether that equipment is a radio or a dishwasher.

We call this a "safety" ground because the purpose is to trip the circuit breaker if voltage appears on the case of your equipment. The safety ground protects you from electrocution.

Do not defeat the safety ground protection with an adapter plug or any other gimmick!

LIGHTNING GROUND

Thor's hammer carries an incredible wallop. A thundercloud can hold over 100 million volts of potential power and typically generates 5,000 to 20,000 amperes of current. A direct hit will be devastating. The lightning ground deflects power from a lightning strike by dissipating the energy before it enters the house.

GROUNDS, LIGHTNING AND EMP PROTECTION

Lightning also creates an electromagnetic pulse. An indirect or nearby hit can induce damaging amounts of current into your antenna and equipment.

Your job is to reduce the potential between, around, and through your equipment and your home's electrical system. Search for and download the Polyphaser "Lightning Protection and Grounding Solutions" PDF book. The book includes robust solutions that commercial installations use to survive direct strikes.

Give lightning a path to the ground outside your home. A lightning arrestor in the feed line and attached to a ground rod is just a start. Multiple 8-foot ground rods, connected in parallel, are better than one.

Lightning arrestors provide some protection against nearby strikes and may bleed off potential charges to prevent a strike. The arrestor has a gap that will allow some of the potential on the conductor to jump to the ground. But, a direct hit will destroy the arrestor. A few nearby strikes can damage the gap. Inspect and replace lightning arrestors often.

Putting an antenna switch in the "ground" position is not enough. The antenna may be "grounded," but all the other cables coming into that switch are connected, and current will flow back, to, and through the connected equipment.

You can't rely on a lightning arrestor, and you can't rely on a switch for full protection. Disconnect your antenna when not in use. I mean completely disconnect, unscrew the cable[24] , and don't leave it lying on any equipment or near a path to ground.

---

[24] You can also find push-on coax connectors for about $5. They are easier to connect and disconnect.

That lightning bolt travels thousands of feet and can easily jump a few more feet to get to your equipment or electrical outlet.

Summer thunderstorms are not the only source of damaging energy. Static electricity due to wind or rain can also cause damage. Dry winter air encourages static charges on the antenna. The components in our receivers handle millionths of a volt not millions of volts. Disconnecting antennas is a good practice during all seasons.

As an additional precaution, unplug your power supply and other equipment to protect against a power surge on the AC line. I lost a very expensive amplifier to a power surge. Even though the amplifier was "off," the power supply was energized. When the lights flickered, the power supply fried.

RF GROUND

We often think of "ground" in terms of direct current or alternating current at house frequencies, 60 Hz. Connecting a wire to your cold water pipe may be a safety ground[25], and it may be limited help to dissipate lightning surges, but it is not an effective RF ground.

"Ground" is not some mysterious pit in which you dump energy. At radio frequencies, every piece of wire has inductance and capacitance. Any wire presents some impedance, which is AC resistance. Therefore, your ground wire may not be at ground potential.

A good RF ground is a low impedance path for RF energy to flow around and not through equipment. If you experience "mic bite" or a burning sensation when

---

[25] No longer recommended, as the water line outside the house could be plastic pipe.

you touch a chassis while transmitting, that is RF. You might also notice distortion on your audio signal. Your computer might reboot for no reason, or the radio display could go haywire.

If you run low power, as we would expect in GMRS service, you might not notice "mic bite" or burning sensations, but you could still have excess RF floating around on the equipment, distorting your signal.

RF can get in your radio equipment in many ways. One is an unbalanced antenna system. Even a balanced antenna such as a dipole can become unbalanced due to objects within its field such as uneven terrain or metal siding. Another cause is feed line that is not perpendicular to the antenna. Balanced feedlines cancel RF, and unbalanced feedlines radiate.

The antenna may induce RF into wires within your house. The house mains, telephones, alarms, televisions, and TV cable all have wires that act as antennas. Also, the various components, such as the computer keyboard, mouse, video display, microphone, coax, and speakers use wire. These wires are antennas. If they pick up the signal from your transmitting antenna, they will add noise and bring RF wherever they are connected. Also, currents on the wires will radiate when you transmit causing RFI in nearby equipment, even though it is not physically connected.

Chokes There are three ways to control RF on wires and feedlines. One is chokes. Chokes do just that, choke off RF. A toroidal choke is a donut of clay and metal. There are different metal mixes chosen for the frequency you are trying to choke. Mix 31 is used below 10 MHz. Above 10 MHz, mix 43 prevails.

Wrap the wire through the choke as many times as will fit. The impedance goes up with the square of the number of turns. If you go through one choke two times, you get four times the impedance. If you go through two chokes once, you only get twice the impedance. More turns are better than more chokes.

Here is an excellent article by K9YC, Jim, dealing with RFI and chokes: audiosystemsgroup.com/RFI-Ham.pdf

Twisted pair The second solution to RF induced in wires is to use twisted-pair wire. The twist will self-cancel any induced currents. You should choose twisted-pair for interconnections.

Bonding The third solution is bonding. Bond each case to the others to provide a low impedance path for RF around the equipment rather than through the equipment. If RF has two paths, one through your equipment and one low impedance around the equipment, most energy will travel on the low impedance path around.

Connect all chassis: radio, antenna tuner, computer, and amplifier with a ground strap to provide a low impedance path from chassis to chassis. That encourages RF to stay off the interconnecting cables and out of the electronics. Also, run a strap to your shack's common ground connection. That is multiple connections between the equipment and station ground. Be sure there is bonding strap parallel to any cables or coax running between components.

RF travels on the surface of a conductor due to a phenomenon known as "skin effect." Ground strap is flat, not round like wire, so it has more surface area. Therefore, strap is a better RF conductor. Skin effect means the strap can be thin. Solid flat strap is a marginally better conductor than braided, but braided strap is more flexible.

To attach the strap to a chassis ground lug, poke a hole through the braid or drill a hole in the solid. Sandwich the strap between a couple of large washers and secure with the nut on your equipment's ground lug. If you need to splice straps, use the same technique, and sandwich them between two large washers.

Finally, electrical codes require you to bond the RF, safety, and lightning grounds together. There should be no difference in potential between the various parts of the grounding system. Any difference in potential will cause current to flow between the components bringing hum, buzz, RFI, or damaging current from EMPs.

The only thing more dangerous than bad grounding is crocodile taunting, as seen here in Chad, North Africa.

ANTENNAS

There is probably no more contentious area of debate than antennas. If you ask five Hams a question, you will get seven different opinions and maybe a fistfight.

With CB, GMRS, and Ham services, you can add an external antenna and enhance your signal. A better antenna helps on both transmit and receive, so money spent on your antenna system is very beneficial. Improving your antenna system is the single most important investment you can make.

There are thousands of antenna designs and hundreds of books and articles on the topic. I am showing you The Easy Way to get on the air without spending a lot of money and with minimal complication. The simple antennas discussed here will fulfill those requirements. You can (and will) spend the rest of your life building, designing and experimenting.

The antenna you have will work lots of stations. The perfect antenna you are looking for, but still haven't installed, won't work anyone.

THE HUMBLE WIRE DIPOLE

We begin with the humble wire dipole. Understanding the dipole is the key to understanding other antennas.

A dipole is a half-wavelength long, cut and fed in the middle. Each side is a one-quarter wavelength long.

You can roll your own very economically. Measure and cut two wires according to the formula 234/frequency = length in feet. Appendix A is a chart for each half of a dipole. That distance is the length of wire from the center insulator to the end insulator. Cut long and trim for results. Removing excess wire is a lot easier than soldering on a longer piece.

ANTENNAS

As you can see from the chart, HF dipoles can be large. VHF and UHF antennas are much shorter and easier to build.

Attach one end of each quarter-wave wire to each side of the center insulator. Feed in the middle by soldering the center conductor to one side and the coax[26] braid to the other. Support the coax on the insulator with a strain relief. The soldered connections can fail if stressed by holding up the weight of the coax. Pre-made center insulators with a standard SO259 coax jack save you the bother of soldering and provide strain relief.

Put another insulator on each end of the wire and attach your hanging rope to the end insulators. UV resistant antenna rope, sometimes called "Para-Cord" is best. Sunlight will destroy other materials in a few months. I have had antenna rope up for over ten years without a problem.

Trim the antenna for minimum SWR,[27] but don't obsess if the SWR is below 1.5:1. Even 2:1 and higher is acceptable if your radio does not start to fold back power. If the SWR goes up with a higher frequency, it means the antenna is too long, wrap a bit of the wire back onto itself. If the SWR goes down with a higher frequency, it means the antenna is too short.

Appendix A is a chart for various frequencies. If your lowest SWR is at 13.9 MHz (202 inches) and you want to move it up to 14.2 MHz (198 inches), the chart tells you to reduce 4 inches from each leg of the dipole. The easy way to do that is to pull 4 inches of wire through the insulator and wrap it around itself at the

---

[26] Coaxial cable. Feedline is in the next chapter.

[27] Standing Wave Ratio is a measure of how well the system is matched. A reading of 1:1 is ideal.

end of the antenna. Shorten an antenna is easy. It is harder to add wire, so start a little long and wrap the excess back.

Your antenna will move up in frequency as you raise it higher, so don't check the SWR on the ground and expect it to stay the same in the air.

Assuming you have a choice, which direction should you hang the antenna? Horizontal wire dipole antennas radiate off their sides, and that is where they have gain (enhanced signal). If your antenna points north/south, it will radiate the best east/west. The pattern for horizontal antennas hung less than a half-wavelength high tends to become omnidirectional, favoring no particular direction.

VERTICALS

The quarter-wave vertical is a cousin of the dipole. It is half a dipole, and something else provides the other half. The other half can be the ground, wire radials, your body holding your HT, or your car body.

Not everyone has trees or other places to hang wire dipoles. Never use a utility power pole as it carries lethal voltages. Perhaps a vertical antenna is your answer.

A properly erected HF vertical can be a good antenna and is relatively stealthy if you are concerned about neighbors and restrictive covenants. Not only is it low profile, but you can also use a tilt-over ground mount and tip the antenna over when not in use — that way the petunia police from the homeowners' association won't notice.

On VHF/UHF, the antennas are short, and you shouldn't have much trouble concealing one. They are also easy to mount on a car using a magnetic base.

ANTENNAS

Mounting the antenna in the center of a vehicle roof provides the most uniform radiation pattern. Mounting your antenna on the corner of a car can distort the pattern making it favor certain directions. Since you are moving, the results are unpredictable. If you have no choice, do what you can.

A particular vertical variation called a Slim Jim, or J-Pole, antenna is cheap and easy to make. They are handy because they are made of transmission line and roll up easily for storage and travel. A GMRS J-Pole is only 18 inches long. Unroll the antenna and hang it on a tree limb as high as you can.

POLARIZATION

The orientation of the electrical field describes a radio wave's polarization. Radiation is the strongest broadside to an antenna. Whichever way the antenna is mounted or facing, most of the energy will be broadside to the wire and not off the ends.

A simple dipole mounted parallel to the Earth's surface is horizontally polarized. The wire is horizontal, and the radiation is the strongest broadside, so the radiation is also horizontal.

A vertical antenna has an electrical field perpendicular to the Earth. Vertical antennas also radiate broadside, and since the antenna is vertical, they are vertically polarized. Mobile FM uses vertical polarization because that is how a car antenna fits and the way you hold an HT antenna - up-and-down. Repeater antennas are vertical as well for the same reason.

If you are using different polarizations, signals are significantly weaker. Don't hold your rubber ducky[28] sideways as that will make the antenna horizontally

---

[28] A "rubber ducky" is the flexible antenna on an HT.

polarized. If you are talking to another person or a vertically polarized repeater, your signal could be as much as 100 times weaker.

When a signal bounces off something, it becomes wobbly like a tumbling hula-hoop. It becomes elliptically polarized (random both ways), and it doesn't matter how your antenna is set up. Skip signals are elliptically polarized, and you can use either a vertically or horizontally polarized antenna.

NVIS ANTENNAS

NVIS stands for Near Vertical Incidence Skywave. A very low HF antenna[29] will direct most of the energy straight up or at a very high angle. When the signal hits the ionosphere, it reflects almost straight back down. NVIS antennas provide good short-range (local to 200 miles) HF coverage. HF using an NVIS antenna will reach out further than your UHF/VHF radio might take you. NVIS on HF is your choice to hit the state capital 100 miles away.

GAIN ANTENNAS

The vertical radiates in all directions. The dipole has some gain (enhancement) off the two sides. A gain antenna concentrates the signal in one direction. The concentration of power can significantly enhance your signal and reception.

For example, a four-element yagi (beam) with 5 watts is about the equivalent of a 40-watt amplifier when compared to a vertical. The yagi is cheaper than an amplifier, has the same multiplier effect on receive, and attenuates signals coming from other directions.

---

[29] How low? Less than half as high as one side of your dipole is long.

ANTENNAS

To make a yagi, start with the humble dipole and add one or more parallel elements. The reflector is about 5% longer than the driven element (dipole). For a three-element beam, add a director about 5% shorter than the driven element to the front of the antenna. The size and spacing of the elements determine the gain, pattern, and bandwidth of the antenna. There are online calculators if you want to roll your own or pick up a commercially made version.

VHF/UHF antennas are small enough they can be hand-held or turned with a small TV rotor. A VHF yagi only has a wingspan of about four feet. A UHF yagi is only about a foot wide.

Suppose you want to establish a fixed link on GMRS between Point A and Point B, but the path a bit sketchy. A gain antenna at one or both ends of the path might solve your problem.

FEEDLINE

A rubber-ducky connects directly to the transceiver. An antenna that is not attached physically to the radio needs a feed line. The most common feed line is coaxial cable or "coax."

Coax consists of a center conductor, a layer of insulation, a shield conductor, and an outer layer of insulation. The dimensions of each determine the impedance,[30] loss, and power-handling capability of the coax.

For impedance, we want to match the impedance of the antenna so the maximum power transfers. For most of our applications, that impedance is 50 ohms, and so is most coax. Almost all coax can handle 100 watts so; your main concern is the loss inherent in the cable.

Loss Loss is measured in decibels (dB) per 100 feet. Decibels express a logarithmic ratio of power. Plus 3 decibels is a doubling of power, plus another 3dB (6 dB) is doubling again or four times the power. Likewise, -3 decibels is half the power, and -6dB is one-quarter the power. If you put 5 watts into a line with a 6dB loss, you get only a little more than a watt at the antenna.

Coax loss goes up rapidly with the frequency and higher SWR, so coax selection is important especially on VHF/UHF. Coax that is suitable for HF, like RG58,[31] has 12dB loss per hundred feet at 400 MHz (UHF). Only one-sixteenth of your power makes it to the antenna. By contrast, LMR400 has a 2.7dB loss and about one-half your power makes the trip.

---

[30] Impedance is resistance to alternating current, in this case, radio waves.

[31] RG58 is a common designation for a popular coax type used on HF. Manufacturers may have their own naming system so consult the charts.

FEEDLINE

Look at a loss chart when selecting coax and go for the lower loss number at your frequency. Keep the lines short. If the loss is 12dB for 100 feet, it will be half that or 6dB for 50 feet.

Moisture Moisture intrusion is the number one enemy of coax. Moisture causes additional losses, and the effect is so gradual you won't notice it over time. If you plan to run your cable underground, get coax rated for that service, or water will penetrate the outer insulation.

Moisture can come in through the connectors. Seal the end of the coax and any connectors to keep out water. Use good quality electrical tape and cover it with Coax-Seal, a putty-like tape. Then cover the Coax-Seal with another layer of electrical tape. The first layer of tape is to make it easier to remove the Coax-Seal later if you need to. If you do it right, coax should last many years.

Finally, moisture can intrude because of animal damage. Squirrels chew on coax and can easily penetrate the outer insulation. I check my lines whenever I can and usually find evidence of squirrel damage. I can't imagine what they find so tasty.

MOBILE OPERATION

CB, GMRS and Ham radios lend themselves to mobile operation. As a bonus, you can connect your home station to your car-mounted antenna and defeat no-antenna regulations. Just don't forget to disconnect before you drive off to work the next day!

Several small transceivers designed for home or mobile operation have a detachable faceplate. Only that small control head goes into the cabin. You mount the transceiver in your trunk or under a seat. Secure the transceiver well, so it does not become a flying missile during a car accident or slide around and create a distraction. Laying a radio on the passenger seat is a bad idea.

Here is an HF/VHF/UHF transceiver in the trunk.

I was fortunate there was a bracket in the spare tire well, and I used it to secure the radio's mobile mount.

Power Run the power cable directly to the battery to reduce electrical noise and provide maximum power. A 100-watt rig draws 15 amps on transmit, so forget about using a cigarette lighter outlet for power. In my case, the battery was also in the trunk making the connection easy. Fuse both the positive and the negative side of the power cable for maximum protection against overload and fire.

Wires Wires for the microphone, speaker, and control cable went under the trunk floor mat, and I removed the back seat to pull them from the trunk to the cabin. An Internet search yielded several videos with complete instructions on how to get that back seat out of the way. Route the cables to the front by tucking them under the door trim for a neat and clean install.

The control head mounted on the dashboard to the right of the steering wheel.

The control head is attached to a bracket and secured with clear Scotch brand "Permanent Double-sided Clear Mounting Tape." It is very sticky but removes without leaving any residue. Some mountings fit in a cup-holder.

The next component was the speaker, and I was at a loss until I looked in the rear-view mirror and noticed the child-seat restraint on the rear deck in the middle of the back seat. Two zip-ties hold the speaker bracket securely to the restraint clip. The wire goes behind the seat and to the trunk where it plugs into the radio.

The speaker looks factory installed.

Before you begin an install, sit in your car and look around. I would never have thought of the speaker mount until it jumped out at me.

Antenna Mobile antennas are one-quarter wave verticals. They may not be a full quarter-wave long, but an inserted coil electrically lengthens the antenna. The car acts as the "other" half of the antenna. Think of the car as radials. You want a good connection to the car body. I use an antenna mount that attaches to the trunk lid. Many trunk hinges ride on fiberglass bearings, so the lid is not bonded to the rest of the car. A strap across the hinge connecting the lid and car body will correct that problem.

MOBILE OPERATION

I call this "the cat's meow."

Dogsled mobile(?) from Greenland.

EXTENDING YOUR RANGE WITH REPEATERS

An Amateur Radio or GMRS HT can talk line-of-sight, a couple of miles on simplex, and much further through a repeater. Simplex communication is the term used to describe an amateur station transmitting and receiving on the same frequency.

We already discussed the fact that VHF and UHF radio waves do not bounce off the ionosphere. They will bounce off buildings, and if you're high enough or conditions are just right for tropospheric ducting, you might be able to communicate further. However, tropospheric ducting is rarely available. We need some help for reliable communication. That's where a repeater comes in handy. [32]

A repeater simultaneously retransmits a signal on a different channel or channels. The repeater "repeats" the signal to relay over a longer distance.

Duplex mode Repeaters operate in "duplex" mode, receiving and transmitting at the same time. It cannot transmit and receive on the same frequency, or the repeater would interfere with itself. The repeater listens and talks on two different frequencies simultaneously to prevent self-interference.

Offset "Repeater offset" is the difference between a repeater's transmit and receive frequencies. A common repeater offset in the two-meter Amateur band (144 MHz) is plus or minus 600 kHz. For

---

[32] Repeaters need power to operate and the premise of this book is "no power." I include the discussion because many repeaters have backup power supplies. Our local repeater is co-located with the county emergency services and we are assured of power as long as they have it. That won't help if the antenna is blown down. Use the repeaters but have another plan when they fail.

example, my local repeater receives on 147.615 MHz and retransmits on 147.015 MHz. To access the repeater, you would tune your receiver to 147.015, so you can hear the repeater, and you would set the offset on your radio to transmit +600 kHz so the repeater can hear you.

On the 440 MHz Amateur band and on GMRS, the repeater split is to transmit 5 MHz higher than the receive frequency. Most GMRS radios do this automatically when tuned to channels 15 through 22, reserved for repeaters.

Height Since VHF/UHF communication is line-of-sight, the higher you are, the further you can see, and the further the radio waves will reach. A repeater antenna will be on a very high site – usually atop a tall building, water tower, or mountain.

In addition to the tall antenna, the transmitter portion of the repeater transmits with more power than the typical handheld or a radio mounted in your car. The combination of extra height and power stretch the repeater's range out to perhaps 50 miles. The useful range will depend on your HT or mobile radio's ability to transmit to the repeater. Those are the basics of a repeater.

Linked network A linked repeater network is a network of repeaters where signals received by one are repeated by the others. Linking greatly expands coverage.

Usually, a local club builds and maintains the repeater with member dues, and that is a good reason to support your local club.

Memory channels A way to enable quick access to a favorite frequency on your ham transceiver is to store the frequency in a memory channel. Your radio will have memory channels, so once you set frequency and offset data in the memory, you do not have to re-enter it every time you turn on your radio or switch to a different repeater. GMRS radios only operate on set channels, so the memory function is built-in.

Tone codes Many repeaters use a sub-audible tone as a "key" for access to prevent interference among repeaters. It is sub-audible, meaning too low in pitch for you to hear. You program the sub-audible tone in a memory channel along with the other repeater data.

Continuous Tone Coded Squelch System (CTCSS) is the term used to describe the use of the sub-audible tone transmitted with normal voice audio to open the squelch of a receiver.

If you're trying to access a repeater and failing, chances are you may not have the proper CTCSS, DTMF, or audio tone for access. DTMF is the tone generated by your keypad. Some repeaters require a DTMF code. Don't let all those fancy names fool you. They are just another way of saying you may need a special tone to unlock the repeater.

A repeater will identify in Morse code every ten minutes. If you can hear a repeater's output but can't access it, a reason might be:

- Improper transceiver offset.
- The repeater may require a CTCSS tone.
- The repeater may require a DCS[33] tone.

APPS for iPhone and Android use the GPS in your phone to locate nearby repeaters and provide access tone information. Look for an APP called "Repeater Book." Write these down or program them in the radio so you are ready when you need them.

---

[33] Digital code squelch.

REPEATERS

Repeaters can also provide a gateway to the internet, and the term "gateway" describes a station used to connect to the internet.

The Internet Radio Linking Project (IRLP) connects Amateur Radio repeater systems via the internet. Repeaters linked by IRLP use DTMF tones generated by your transceiver keypad just like the tones on your telephone. DTMF stands for Dual Tone Multi-Frequency. Dial-in the "phone number" of a distant repeater and your signal will pop out of that repeater – even on the other side of the world. Dial 5600 and your signal will travel over the internet to a repeater in London.

Echolink is another internet linking protocol.

Repeaters are backup devices Repeaters are a back-up not a primary mode for emergency services. Our scenario assumes no power. Even if there is power, the repeater antenna can blow down. The repeater may suffer damage from lightning, EMP or other misfortune. Repeaters can fail so you should be prepared to operate in simplex mode. This requires planning, practice and coordination. There is more in the chapter on Operating Tips and Strategies.

MESH NETWORKING

Mesh networking is an application that repurposes wireless routers to create a network over Amateur Radio. The nodes interconnect in a spider web fashion and route the message along whatever path it might need to get to the final destination. If one node drops out, the software sends the message through other nodes to reach home. The internet is essentially a mesh network. Amateur operators are creating a wireless internet.

Wireless computer routers share frequencies with a portion of the 2.4, 3.4, and 5.8 Gigahertz (GHz) Ham bands. Some industrious Hams, exercising what I call "hamgenuity," saw to modify a router's firmware to adapt it to mesh networking. They also boosted the router output. Hams can attach amplifiers and gain antennas greatly increasing the range of an individual node.

Mesh networking is relatively inexpensive ($150 per node) and can provide a very robust data network. A combination of fixed and portable stations could cover a wide local area. Users can configure nodes to connect to the internet as well.

Anything "data" can travel over a mesh network, including email, video, or voice-over-internet (Skype) audio.

One group spearheading the effort is the Amateur Radio Emergency Data Network (AREDN). They define their responsibility as, "The AREDN™ development team strives to create quality software releases for use on commercial-off-the-shelf (COTS) devices with a primary focus on meeting the needs of emergency communications data networks." The website is AREDN.org.

MESH NETWORKING

Another group promoting mesh networking has a website, Broadband-Hamnet.org. Their product is "a high speed, self-discovering, self-configuring, fault-tolerant, wireless computer network that can run for days from a fully charged car battery, or indefinitely with the addition of a modest solar array or other supplemental power source. The focus is on emergency communications."

It helps to have a computer networking background when setting up a mesh network, but both groups are working to make the system more plug-and-play.

Mesh networks show great promise for groups wishing to set up a wide-area wireless network designed to operate without mains power.

OPERATING TIPS AND STRATEGIES

You are preparing a communications plan, and that may require multiple approaches. CB, GMRS, and Ham VHF/UHF are the obvious choices for local contacts. For longer distances, you need HF. There are millions of permutations of time and frequency. You should arrange a meeting place (frequency) and time for your inner circle.

COMMUNICATION PLANNING

Before Titanic sank, there was no generally accepted radio protocol. The US and Europe heard Titanic's distress signals, but the nearest ship's operator had turned off his equipment and gone to bed. International reforms established designated listening times when all transmitting stopped, and ships would listen for distress calls. Preppers need a similar system, and one is the 3-3-3 rule.

3-3-3 Rule The 3-3-3 rule is simple:

- Turn on your radio every 3 hours,
- Listen for at least 3 minutes,
- On channel 3 or the designated ham frequencies.
- Follow the same schedule to call for help if you need it.

The start time is noon, so you listen at noon, 3 PM. 6 PM, etc. You listen for at least three minutes on channel 3 of FRS, MURS, or CB. Appendix F lists designated frequencies.

Your inner-circle could choose a different protocol, but the important point is to have a plan so people can reasonably expect to be on the same frequency at the same time.

OPERATING TIPS AND STRATEGIES

The advantages of this plan are:

- Easy to remember.
- Short listening periods conserve battery life.
- Everyone is on the air at the same time and frequency.
- Sets a schedule of 8 times a day to call
- You don't have to call for hours, hoping someone will tune in to hear you.
- The 3-hour schedule gives time to change locations for better reception.

The 3-3-3 rule only consumes a maximum of 24 minutes listening each day. If your battery is good for ten hours, it will last three weeks.

PHONETICS

Standard phonetics help cut through noise and interference.

Here is the word list adopted by the International Telecommunication Union:

A--Alpha
B--Bravo
C--Charlie
D--Delta
E--Echo
F--Foxtrot
G--Golf
H--Hotel
I--India
J--Juliett
K--Kilo
L--Lima
M--Mike
N--November
O--Oscar
P--Papa
Q--Quebec
R—Romeo
S--Sierra
T--Tango
U--Uniform
V--Victor
W--Whiskey
X--X-ray
Y--Yankee
Z—Zulu

The FCC originally assigned me the call sign KG4CVN. "G," "C" and "V" sound alike especially in noise or fading conditions. So do "N" and "M." Golf, Charlie, and Victor don't sound anything alike. No one can confuse November and Mike. Using standard phonetics makes it easier for others to understand.

Notice I said "standard phonetics." Cutesy or non-standard variations can only confuse.

HOW TO USE A MICROPHONE

You may think this is elementary, but if you listen around, you will hear plenty of poor mic skills. Holding the mic too far away results in lots of background noise. Talking too quietly doesn't "fill" the signal and is hard to understand. Yelling into the mic causes distortion and strains your voice.

When you call, be clear and crisp. Don't mumble, don't shout and don't stretch out your phonetics. Keeeeellllloooo is not easier to understand than Kilo.

Slow down and enunciate clearly. The message will get through quicker if you don't have to repeat yourself three times. Avoid confusing fillers such as "err' and "uh."

Don't yell into the microphone. It won't make your signal any stronger but will produce distortion. Get close to the microphone and speak in a normal tone of voice. Practice and ask other people how you sound. You may need to play with the microphone gain and compressions controls, if there are any.

ENCRYPTION AND CIPHERS

Laws prohibit the use of ciphers or codes to obscure a message. In licensed services, if you speak a foreign language, you are to identify in English. You can develop your own trigger words. For instance, "The cow has left the barn." could mean, "I am on my way." In a real crisis, you will do whatever you have to do, but when practicing in peacetime, use real English.

Don't be an elephant, stomping over everyone attempting to make contact.

United Nations Headquarters Station, New York

ANATOMY OF AN AMATEUR RADIO HF CONTACT

Always check the frequency before you transmit. Be sure you are within limits for your license class. You will hear Extra Class licensees in the lower portion of the band, and you will hear foreign stations on 40 meters conversing where no US phone station can operate. Don't assume because you hear voices that you can transmit there as well.

Make sure the frequency is clear and not already in use. Ask, "Is the frequency in use?" and identify yourself.

What is the right way to start and sustain a conversation? You could start by calling "CQ." CQ is a general call to anyone, inviting him to join you. Never use CQ on a repeater. Just announce your call sign. "K4IA listening."

QSL card from Germany.

ANATOMY OF AN HF CONTACT

While you are new, I don't recommend you call CQ until you get familiar and comfortable with conversing on the radio. Listen and find someone else calling CQ. Why?

Ham etiquette says the person who called CQ sets the tone and pace of the conversation. Admit it; you don't know what you are doing. Mr. CQ goes first. He gives you a signal report, location, and name. You give him yours. He might describe his equipment. You tell him about yours. Let Mr. CQ lead. You don't need to panic about, "What do I say next?" Follow the other guy's prompt and respond to him. Try to think of a question to keep the conversation going.

Keep your transmissions short. Regularly spoken conversation is one or two sentences, and then the other person speaks or interrupts. If you are transmitting, you can't hear the other side. You could go on forever, as no one can interrupt. Long discourses are tough to follow. Don't be a hog. The other person should not have to take notes, so he remembers your 14 points before responding.

Exchanges are more interesting if they proceed like a regular conversation. The pattern should be brief two-ways and not extended monologs.

You answer a CQ by giving his call sign once and yours twice. Use phonetics with your call sign. "WA4TUF this is Kilo 4 India Alpha, Kilo 4 India Alpha." You might add, "How copy?" "Over." or say nothing. Mr. CQ will hear the silence and know you have stopped calling.

A conversation does not require identifying on every "over." A simple, "What do you think, John?" is enough to let the other guy know it is his turn. You do identify every ten minutes, and at the end of your contact as the Amateur rules require.

Once you get comfortable with HF conversations and believe yourself ready to lead one, then you can call CQ. Before you call CQ, listen. Make sure the frequency is not in use. Sometimes, you can't hear both sides talking, so you should identify and ask, "This is K4IA. Is the frequency in use?" Listen for a few seconds and ask again.

When you're sure you are on a clear frequency appropriate for your license, you can make your call. "CQ CQ CQ this is K4IA, Kilo 4 India Alpha, Kilo 4 India Alpha, calling CQ and listening."

I prefer several short calls to one long one. Listeners will tire of a long CQ and spin the dial. If my short CQ doesn't get an answer, I call again after waiting at least five seconds.

When you get an answer, acknowledge the station, so he knows you are calling. "WA4TUF, K4IA returning. Thanks for the call. You're 58[34] in Fredericksburg, Virginia. Name here is Buck, Bravo Uniform Charlie Kilo. How copy?"

Notice, I use his call once. He knows his call sign and presumably will recognize it. I don't need to repeat my call. He knows it because he called me. I might repeat the signal report if conditions are poor but usually once is enough. I don't spell out Fredericksburg phonetically. It is long enough as it is. If the other guy doesn't get it the first time, he can ask. I do spell out my name, maybe twice if conditions are poor or the other guy sounds weak.

That initial exchange is the classic trinity of signal report, location, and name. Where it goes from there is up to you. On the next go-arounds, I might tell him

---

[34] A Signal Report. 1-5 for overall readability and 1-9 for signal strength.

where Fredericksburg is, my equipment, my weather, my other hobbies, what I do for a living, how long I have been a Ham, or my favorite Ham Radio things to do. That would take several transmissions.

So far, I have only talked about me. If you want to be a good conversationalist, learn to listen. Ask the other guy about things that interest him. "My rig is a WizBang 4000 Pro. What's yours?" "I like to garden. What do you do when you're not on the radio?" You will find common ground or learn something new.

Eventually, you will run out of things to say. How do you end the conversation? You could take the coward's way out and claim the XYL[35] is calling, the dog is barking or the phone rang. There is no need for subterfuge. It is acceptable to say, "Thanks for the contact, John. I think I'll try to scare up a few more before I go to bed. Hope to hear you on again soon. 73 and good night. WA4TUF this is K4IA clear on your final." Adding, "Clear on your final'" lets listeners know you are not quite done and discourages tail-ending interrupters.

Ending a radio conversation is not nearly as awkward as walking away from a cocktail party windbag.

You will meet some very intriguing people. I've included some interesting QSL cards to prove the point. I have shoeboxes full and enjoy exchanging them with Hams around the world.

Take "Tex" on the next page. Tex, W5 Big, Quick, Ugly was 101 when I talked to him on my commute home. He was quite a character.

---

[35] XYL is Hamspeak for wife.

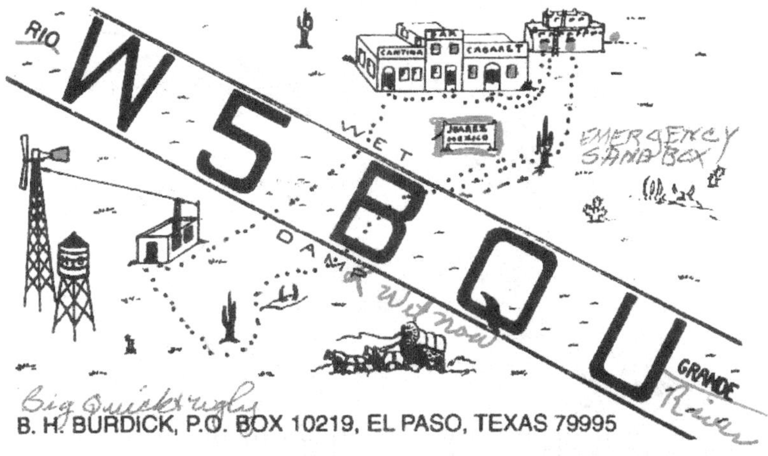

At the time we talked, Tex was reputed to be America's oldest active Ham. He is a Silent Key[36] now.

Kay was ARRL President when we met on the air.

[36] Silent Key refers to a deceased Ham. It is a throwback to the days of telegraph.

Caveman portable

WAITING FOR AMERICA AND GETTING CHINA INSTEAD!

Who can resist this card from Northern Ireland?

A22 is the prefix for Botswana. A Japanese Ham was operating there. Hence, the call sign A22/JA4ATV.

Greetings from the south of Africa

Artur Makhtsiev, ex UA6JFF, RA6JF

☑ ZS6BQI ☐ 7P8BA

Joburg, Republic of South Africa Maseru, Lesotho

Radio	DD/MM/YY	UTC	MHz	Two way	RST
K4IA	22.12.07	1848	18	CW	549
	10.01.09	2101	10,1	CW	579
via:					

RV1CC prdnt

☐ PSE QSL TKS ☑ via buro or:

WAZ: 38 ITU: 57

P.O.Box 44698, Linden 2104, South Africa

73! Art

ZS is South Africa. Art is licensed there and in Lesotho.

One of my favorite cards is from Kazakhstan.

Gyuri is QRL (busy) chasing QSL cards.

You can talk to the radio room of the *HMS Queen Mary, USS Wisconsin*, and many other vessels. The *USS Wisconsin* is berthed in Norfolk, Virginia.

"Big Mo" is stationed at Pearl Harbor, Hawaii. The Japanese surrender ending WWII took place on its decks.

ANATOMY OF A REPEATER CONTACT

Just like HF, first, check to be sure that the frequency is not in use. Listen before you leap. Silence does not mean the repeater is not in use. There may be many stations in standby mode during an emergency.

You do not call CQ on a repeater. If you want to invite a general call, you say your call sign. "K4IA listening." If you want to initiate contact with a specific station, call him. "WA4TUF, this is K4IA, you out there?"

Don't make any unanswered call more than twice without waiting a few minutes. If there is anyone out there who wants to talk, they will answer.

Most repeater users accept your joining in an existing conversation. To do that, wait for an appropriate break and say your call sign. You'll be recognized and invited to join. Consider the conversation before you call. If the participants are deep in a discussion of antenna theory, don't interrupt to comment on something off-topic. That's rude.

"Break Break" is reserved for emergencies only. Don't use it to join a conversation.

Repeater transmissions are generally short. The repeater will time-out and drop your transmission if you go on too long (usually defined as more than a minute). This is to allow another caller to jump in. It also protects against the dreaded "open microphone," when someone sits on their microphone and engages the push to talk switch by mistake. Yes, that happens.

Pause between transmissions to listen for someone who might want to join your conversation or report an emergency.

Identify with your call sign every ten minutes and at the end of your conversation. The interval is fifteen minutes for GMRS transmissions.

NETS

A net is a group of users. Informal nets are roundtables, where the microphone is passed around, in order: Joe to Moe to Larry. Everyone gets a turn and passes to the next participant.

A formal net is "directed." There is a net control station (NCS) that manages the session. He would first call the net to order, reciting a monolog of the net's purpose and protocols. Then, the NCS will ask for check-ins. First, come stations with emergency traffic (messages), then mobile stations, and then general check-ins. The NCS will recognize the individual stations and ask for their comments, one at a time. If two stations need to converse, they will be sent off the net frequency so as not to tie up the net and the frequency.

If you hear a net operating, you can check in but should standby until called on or you hear instructions.

TROUBLESHOOTING

In the scientific method, parsimony is an epistemological, metaphysical, or heuristic preference, not an irrefutable principle of logic or a scientific result. Occam's razor would demand that scientists accept the simplest possible theoretical explanation for existing data.

Troubleshooting radios follows the same principle. The problem is usually something simple. That is a good thing because modern radios are almost impossible to work on without training and sophisticated test equipment. Radio parts are minuscule and mounted on layers of stacked circuit boards. Stay out of the radio unless you know what you are doing.

As a rule, you start troubleshooting by asking, "What changed?" Did you recently change power sources, antennas, microphones, etc. Go back and undo the change. Did you recently relocate the equipment? Perhaps a cable is disconnected. Was there a storm that could have blown down your antenna?

What can go wrong? Most problems are operator error.

EQUIPMENT FAILURE

If the radio doesn't turn on, check the power and associated cables:

- Is the power source "on?"
- Check the power connectors to make sure one isn't loose or reversed.
- Use a multi-meter to measure the voltage on the power source. It should read around 14 volts.
- If the power source is good, check for blown fuses. A blown fuse may show blackened glass, and you won't be able to see the thin wire that

normally connects the two ends. If you can't tell, use a multi-meter to check the resistance from one end of the fuse to the other (with the fuse removed).

- If the resistance is infinite, you have a blown (open) fuse. Try to deduce why it blew. Something caused the radio to draw excessive current. Did you cross the power wires? Are you operating into a badly mismatched antenna? Put in another fuse of the same size, and hope it was just a fluke. Don't upsize the fuse, or you risk equipment failure or fire.
- If the fuse is good, check the voltage at the radio end of the cable.
- If there is no voltage at the radio, disconnect the power and check the resistance of the power cable. If it is infinite, the cable or connector is broken and needs to be patched or replaced. Usually, it is the connector and specifically where the wire attaches to the connector.

I CAN'T HEAR ANYBODY OR THEY CAN'T HEAR ME

If the radio turns on, but you don't hear any signals:

- Is the radio locked in transmit mode?
- Is the RF or audio gain turned down?
- Is the squelch control turned up too high for signals to trigger?
- Is the antenna connection loose? Unscrew it and re-attach to be sure.
- You could have a faulty antenna connector.
- Measure the antenna feed line with a multi-meter. Set it to ohms. There should be no resistance along the cable and infinite resistance between the core and the shield. Some antennas have a coil across the core and ground. They will exhibit a short when

connected to an ohm-meter. Best to disconnect the antenna from the feedline when measuring.

- Try a different antenna.
- If you are using a repeater, do you have the offset and tones set correctly? This is usually the problem.
- If you are using an FRS or GMRS radio, do you have the tones set correctly?

If you can hear signals, but no one hears you transmitting:

- Is the radio in the wrong mode? The microphone doesn't work if the radio is in a non-voice mode like CW or Digital.
- Is the microphone gain turned down?
- Is your microphone cable frayed, loose or not attached properly?
- Attach an SWR or power meter and see if it shows any signal going out.
- Ask a friend to listen for you.
- If you are using a repeater, FRS or GMRS radio make sure you have the offset and tones set correctly.
- Check that your radio is not in low power, test, or standby mode.
- Are you using a properly sized power supply wire? If the wire is too long or too thin, voltage drop could affect the transmitter. The receiver sounds fine because it draws little current and has less voltage drop.

LEARNING MORSE CODE

If I told you I was going to teach you a foreign language and all you needed to know was 26 words and how to count to 10, would you think it an impossible task? Probably not.

Boy Scouts was my first experience with Morse code. Unfortunately, I learned it all wrong and had to relearn it as a Ham. What went wrong?

In Boy Scouts, we memorized Morse as dots and dashes. The first exposure was visual, not audible. You had a little chart and looked up every letter. What's wrong with that? Your brain has to translate what it sees or hears into dots and dashes and then translate those into letters. It becomes a multi-step process. It is like learning to translate from English to Spanish to get to French.

I thought I understood it pretty well until someone sent Morse code by flashlight. It was blinking light to dots and dashes to letters. It was even worse when sent by waving a flag. I was lost.

Radio Morse is audible, and you must learn to recognize the sound and not memorize dots and dashes. "A" is not dot-dash. It is not even dit-dah. It is the sound made by dit-dah. Learning the sound eliminates all the in-between translations.

For the same reason, do not learn that "A" sounds like "Ah-pull" or that the letter "A" has a short line and a long line. I've seen pictograms like children's alphabet blocks. These gimmicks introduce additional mental steps. Now you are going from English to Spanish to French to get to Russian.

LEARNING MORSE CODE

Koch method Learn Morse by hearing it one or two letters at a time until you can immediately make the connection from the sound to the letter. Then you go on to the next letters and build. This is called the Koch method.

Another impediment to my learning was the way we sent Morse. At slow speeds, "A" became the sound made by diiiiiit-daaaaaaaaaah. Then, the next letter came immediately with no time for the brain to work.

Farnsworth method Farnsworth timing is the modern way and sends the letters at least 20 words per minute,[37] so each letter has one distinct sound. Dits and dahs may form a letter, but you should not try to hear individual elements. The letter is one sound.

To slow down the sending pace, Farnsworth increases the space between letters and words. Increasing space does two things. It reinforces a single sound as the letter, and it gives the brain extra time between letters and words to do the translation. You should also send using Farnsworth timing because that is how the other guy learned as well. I set my keyer around 22-26 words per minute and slow down by increasing spaces.

After learning the letters, you will start to recognize words. When you read, you do not see letters; your brain jumps to the word. "Word" is not "W-O-R-D." The same happens with Morse Code. You learn to recognize your call sign, RST, 5NN, TU, 73, and other common "words" without thinking about the individual letters or the elements that make up the letters.

---

[37] Words per minute (WPM) is based on a 5 letter word. "Paris" is an often-used standard. Send "Paris" 20 times in a minute for 20 WPM.

It all takes practice. In the beginning, listen to a code practice CD or audio file. K7QO offers a free course download on his website, K7QO.net. G4FON has a Koch trainer at G4FON.net. There are other sources, as well.

Once you get the letters down (mostly), listen to QSOs[38] for the standard QSO pattern. Then, GET ON THE AIR! There is no better practice than making actual QSOs. Real QSOs are exciting and won't seem like practice.

Getting on the air is the best way to practice.

FISTS CW Club and Straight Key Century Club (SKCC) promote CW, and you will find slow CW on the frequencies suggested on their web pages. Both FISTS CW Club and SKCC assign you a member number that you exchange with other members to collect awards. It is a fun challenge. Suggested frequencies to find slower CW include:

[38] QSO is a contact or conversation.

3.550 - 3.600 MHz
7.055 - 7.125 MHz
14.055 - 14.060 MHz
21.055 - 21.200 MHz
28.055 - 28.060 MHz

FISTS USA
FISTS Number 10,000
KNOWCW
USA Club Call of FISTS CW Club
Headquarters: PO BOX 47, Hadley MI 48440

Radio	Date	UTC	MHz	2-X	RST	Power	QSL
K4IA	5/13/06	1758	7MHz	CW	599	100 W	Tnx Pse

FIST members can get permission to use the FISTS CW Club call sign.

When starting out on CW, concentrate on the QSO trinity of signal report, location, and name first. The pattern is always the same, so you know what to expect. With time, you will progress to more complex conversations. Try to match the speed of the other guy but if you can't, "QRS" means "slow down" and "QRQ" means "speed up."

For more serious practice, tune to the W1AW Code Practice Sessions. These texts from *QST* magazine are harder to copy because the words are longer and not as predictable. ARRL offers code proficiency certificates that will look impressive on your wall. http://www.arrl.org/w1aw-operating-schedule

Greetings from QRP Station

K5QLF

Fred Bonavita
334 Royal Oaks Drive
San Antonio, Texas 78209
Bexar County

Power is no substitute for skill

Ex: W5QJM, W7JLX, W4WUQ, ZF2AL
G-QRP 819; QRP ARCI 4577; Nor Cal 69
CW Operators QRP Club 31; FISTS 3231
SMIRK 3476 (EL09) CC = 990

To: K4 IA
OP: BUCK

When someone has a lousy fist[39], you joke, "QLF? Are you sending with your left foot?"

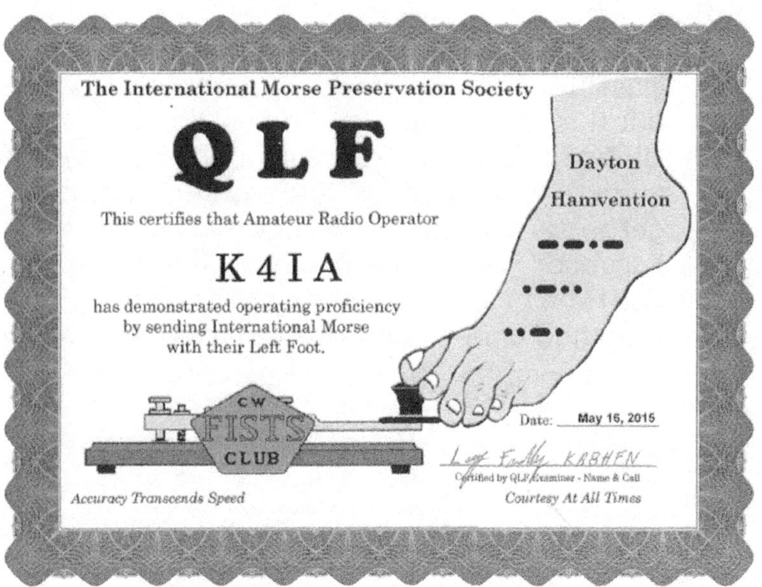
The International Morse Preservation Society

QLF

This certifies that Amateur Radio Operator

K4IA

has demonstrated operating proficiency by sending International Morse with their Left Foot.

Dayton Hamvention

Date: May 16, 2015

Certified by QLF Examiner - Name & Call: KA8HFN

CW FISTS CLUB

Accuracy Transcends Speed

Courtesy At All Times

I earned this certificate for demonstrating proficiency while sending with my left foot.

39 "Fist" refers to the operator's style of sending; his "accent" when sending CW.

AMATEUR RADIO RESOURCES

ARRL The American Radio Relay League has an old-fashioned name. It has been around since 1914. The League is a prolific publisher, advocate, and defender of Amateur Radio, billing itself as "the voice of Amateur Radio." The website, ARRL.org, is full of excellent articles and advice.

Magazines The two preeminent Ham Radio print magazines are *QST* and *CQ*. I enjoy them both. Both are available in paper or digital versions.

- *QST* is the official journal of the ARRL and comes as a benefit of membership in the American Radio Relay League ($49/year). ARRL.org/membership
- *CQ* is less formal and less technical but more fun. CQ-Amateur-Radio.com ($27/year)

Online

- eHam.net is full of articles, equipment reviews, classified ads, and news. The material is free, but they encourage you to contribute.
- QRZ.com is also a free resource with news and an extensive call sign database. Subscribers get to see more information.
- Facebook has groups for every interest.
- YouTube has videos of almost everything, including unboxings, reviews and "how-tos."

SUMMARY AND RECOMMENDATIONS

The purpose of this book was to show you The Easy Way to prepare yourself for emergency and prepper communications. Here is a summary of The Easy Way methods:

- Get involved with a local radio club.
- Find an Elmer or two.
- Listen, listen, listen, and learn.
- Get on the air no matter how modest your station or antenna.
- Don't despise humble beginnings.

My recommendations for a station would be:

- AM/FM/Shortwave/NOAA portable radio.
- CB and GMRS radio for local communication.
- At least three power sources for all equipment.
- Inverter type generator.
- Battery backup (20 amp-hour or more) with at least 100 watts of solar cells.
- Amateur Radio Technician License.
- Portable HT and more powerful base station, Amateur or GMRS.
- Directional antenna and/or a way to elevate your antenna at least 15 feet.
- Amateur Radio General Class license.
- HF station.

73/DX
Buck
K4ia

APPENDIX A – ANTENNA LENGTHS

Values are for each half of a dipole based on 234/frequency = length in feet. Cut longer and adjust. To adjust, note the actual and target frequency of lowest SWR. Add or subtract the difference in length on each half of your antenna. To move lower, add wire. To move higher wrap it back. Bold frequencies are the approximate Amateur bands.

MHz	Feet	Inches	MHz	Feet	Inches
	80 M			**30 M**	
3.2	73.2	878	10	23.4	281
3.3	71	851	10.05	23.3	279
3.4	68 5/6	826	**10.1**	23.2	278
3.5	66 6/7	802	**10.15**	23	277
3.6	65	780	10.2	23	275
3.7	63.25	759	10.25	22.8	274
3.8	61.6	739	10.3	22.7	273
3.9	60	720	10.35	22.6	271
4.0	58.5	702	10.4	22.5	270
4.1	57	685	10.45	22.4	269
4.2	55 .7	669	10.5	22.3	267
	40 M			**20 M**	
6.7	35	419	13.9	15.8	202
6.8	34.4	413	**14.0**	16.7	201
6.9	34	407	**14.1**	16.6	199
7.0	33 3/7	401	**14.2**	16.5	198
7.1	33	395	**14.3**	16.3	196
7.2	32.5	390	**14.4**	16.25	195
7.3	32	385	14.5	16.1	194
7.4	31 5/8	379	14.6	16	192
7.5	31.2	374	14.7	16	191
7.6	30.8	369	14.8	15.8	190
7.7	30.4	365	14.9	15.7	188

MHz	Feet	Inches	MHz	Feet	Inches
	17 M			**10 M**	
17	13.75	165	27.5	8.5	102
17.5	13.4	160	28	8.3	100
18	13	156	**28.2**	8.3	100
18.068	13	155	**28.4**	8.25	99
18.168	12.8	155	**28.6**	8.2	98
18.25	12.8	154	**28.8**	8.1	98
18.35	12.75	153	**29**	8	97
18.45	12.6	152	**29.2**	8	96
18.55	12.6	151	**29.4**	8	96
18.65	12.6	151	**29.6**	8	96
18.75	12.5	150	30	7.8	94
	15 M				
20.8	11.25	135		**VHF**	
20.9	11.2	134	**144**	1.625	20
21.0	11.1	134	**148**	1.6	19
21.1	11	133			
21.2	11	132		**UHF**	
21.3	11	132	**440**	.5	6
21.45	11	131	**448**	.5	6
21.55	10.8	130			
21.65	10.8	130		**GMRS**	
21.75	10.75	129	462	.5	6
21.85	10.7	129	467	.5	6
	12 M				
24	9.75	117			
24.849	9.4	113			
24.94	9.4	113			
25.1	9.3	112			
25.2	9.3	111			
25.3	9.25	111			

APPENDIX B - ABBREVIATIONS AND PROSIGNS* FOR MORSE CODE

ABT	about
AGN	again
ANT	antenna
AR*	end of message
B4	before
BCNU	be seeing you
BK*	break (sent as one letter)
BTR	better
BURO	bureau (QSL buro)
C	yes
CFM	confirm
CK	check
CPY	copy
CUD	could
CUAGN	see you again
CUL	see you later
DE	this is
DR	dear
ENUF	enough
ES	and
FB	fine business
GB	good bye
GM, GA, GE, GN	good morning etc.
GG	going
GUD	good
HI HI	☺ LOL laughter
HR	here or hear
HV	have
HW	how
KN*	no one else answer
LID	poor operator
LIL	little
N	nine (RST 5NN = RST 599)
OM, OT, OC, OB	old man, top, chap, boy
OP	operator or name

PSE	please
PWR	power
R	roger, OK
RCVR, RX	receiver
RIG	equipment
RPT	repeat or report
SASE	self-addressed stamped envelope
SED	said
SHD	should
SIG	signal
SK*	end of conversation
SKED	schedule
T	zero (pwr 1TT w = 100 watts)
TFC	traffic
TMW	tomorrow
TT	that
TNX,TKS	thanks
TU	thank you
TX, XMTR	transmitter
U, UR, URS	you, your, yours
WID	with
WK, WKD	work, worked
WL	will
W,WTS	watts
WUD	would
WX	weather
XCVR	transceiver
XTAL	crystal
XYL	wife
YL	young lady
73	best regards
88	love & kisses

*ProSigns are sent as one letter

APPENDIX C - Q SIGNALS

Q signals are the Amateur Radio equivalent of 10 signals used by CB and police. They become a question when followed by a question mark. QTH? Where are you located? Answer: QTH VA.

QRG Frequency.

QRL Is the frequency in use?

QRM I am troubled by interference.

QRN There is noise / static.

QRO Increase power.

QRP Decrease power or low power.

QRQ Send faster.

QRS Send slower.

QRT Stop sending or I am going off the air.

QRU I have nothing further for you.

QRV I am ready.

QRX Standby. QRX 10. Standby 10 minutes.

QRZ Who is calling me?

QSB Signals are fading up and down.

QSK I can hear you between my signals. You can break in during my transmission.

QSL Acknowledge receipt.

QSO Contact.

QSY Change to another frequency.

QSX Listening frequency

QTH Location.

APPENDIX D - 50,000 WATT AM RADIO STATIONS

The most power allowed on the AM broadcast band in the United States is 50,000 watts. You can often hear these mega-stations at night, coast-to-coast. There is no guarantee they will operate at full power in an emergency, but you may be able to get information from a part of the country not affected by your local conditions.

The list is in ascending order by frequency.

WFLF	540 kHz	Orlando, Florida
KMJ	580 kHz	Fresno, California
WTCM	580 kHz	Traverse City, Michigan
WTMJ	620 kHz	Milwaukee, Wisconsin
WGST	640 kHz	Atlanta, Georgia
KFI	640 kHz	Los Angeles, California
WWJZ	640 kHz	Mount Holly, New Jersey
WNNZ	640 kHz	Westfield, Massachusetts
KENI	650 kHz	Anchorage, Alaska
WSM	650 kHz	Nashville, Tennessee
WLFJ	660 kHz	Greenville, South Carolina
WFAN	660 kHz	New York
KTNN	660 kHz	Window Rock, Arizona, Navajo Nation
KBOI	670 kHz	Boise, Idaho
WSCR	670 kHz	Chicago, Illinois
KLTT	670 kHz	Commerce City, Colorado
WWFE	670 kHz	Miami, Florida
WCBM	680 kHz	Baltimore, Maryland
WRKO	680 kHz	Boston, Massachusetts
WCNN	680 kHz	North Atlanta, Georgia
WPTF	680 kHz	Raleigh, North Carolina
KKYX	680 kHz	San Antonio, Texas

50,000 WATT AM STATIONS

KNBR	680 kHz	San Francisco, California
WJOX	690 kHz	Birmingham, Alabama
WOKV	690 kHz	Jacksonville, Florida
WLW	700 kHz	Cincinnati, Ohio
KALL	700 kHz	North Salt Lake City, Utah
KXMR	710 kHz	Bismark, North Dakota
KSPN	710 kHz	Los Angeles, California
WAQI	710 kHz	Miami, Florida
WOR	710 kHz	New York
KIRO	710 kHz	Seattle, Washington
KEEL	710 kHz	Shreveport, Louisiana
WGN	720 kHz	Chicago, Illinois
KDWN	720 kHz	Las Vegas, Nevada
WGCR	720 kHz	Pisgah Forest, North Carolina
KBRT	740 kHz	Costa Mesa, California
KNFL	740 kHz	Fargo, North Dakota
KTRH	740 kHz	Houston, Texas
WYGM	740 kHz	Orlando, Florida
KCBS	740 kHz	San Francisco, California
KRMG	740 kHz	Tulsa, Oklahoma
KFQD	750 kHz	Anchorage, Alaska
WSB	750 kHz	Atlanta, Georgia
KERR	750 kHz	Polson, Montana
KXTG	750 kHz	Portland, Oregon
WJR	760 kHz	Detroit, Michigan
KTKR	760 kHz	San Antonio, Texas
KFMB	760 kHz	San Diego, California
KDSP	760 kHz	Thornton, Colorado
KKOB	770 kHz	Albuquerque, New Mexico
KCBC	770 kHz	Manteca, California
WABC	770 kHz	New York
KTTH	770 kHz	Seattle, Washington
WBBM	780 kHz	Chicago, Illinois
KKOH	780 kHz	Reno, Nevada
WHB	810 kHz	Kansas City, Missouri

WSJC	810 kHz	Magee, Mississippi
KGO	810 kHz	San Francisco, California
WKVM	810 kHz	San Juan, Puerto Rico
WGY	810 kHz	Schenectady, New York
KGNW	820 kHz	Burien, Washington
WBAP	820 kHz	Ft Worth, Texas
KUTR	820 kHz	Taylorsville, Utah
WTRU	830 kHz	Kernersville, North Carolina
WCCO	830 kHz	Minneapolis, Minnesota
KLAA	830 kHz	Orange, California
KFLT	830 kHz	Tucson, Arizona
WCRN	830 kHz	Worcester, Massachusetts
WCEO	840 kHz	Columbia, South Carolina
WHAS	840 kHz	Louisville, Kentucky
KXNT	840 kHz	North Las Vegas, Nevada
WXJC	850 kHz	Birmingham, Alabama
WEEI	850 kHz	Boston, Massachusetts
WKNR	850 kHz	Cleveland, Ohio
KOA	850 kHz	Denver, Colorado
KICY	850 kHz	Nome, Alaska
WTAR	850 kHz	Norfolk, Virginia
KTRB	860 kHz	San Francisco, California
KPAM	860 kHz	Troutdale, Oregon
KRLA	870 kHz	Glendale, California
WWL	870 kHz	New Orleans, Louisiana
KRVN	880 kHz	Lexington, Nebraska
WCBS	880 kHz	New York
KLRG	880 kHz	Sheridan, Arkansas
WBAJ	890 kHz	Blythewood, South Carolina
WLS	890 kHz	Chicago, Illinois
KTIS	900 kHz	Minneapolis, Minnesota
WFDF	910 kHz	Farmington Hills, Michigan
KFIG	940 kHz	Fresno, California
WMAC	940 kHz	Macon, Georgia
WINZ	940 kHz	Miami, Florida

50,000 WATT AM STATIONS

WWJ	950 kHz	Detroit, Michigan
KJR	950 kHz	Seattle, Washington
WTEM	980 kHz	Washington, D.C.
WDYZ	990 kHz	Orlando, Florida
WNTP	990 kHz	Philadelphia, Pennsylvania
WMVP	1000 kHz	Chicago, Illinois
KOMO	1000 kHz	Seattle, Washington
WTZA	1010 kHz	Atlanta, Georgia
WJXL	1010 kHz	Jacksonville Beach, Florida
WINS	1010 kHz	New York
WHFS	1010 kHz	Seffner, Florida
KXEN	1010 kHz	St. Louis, Missouri
KIHU	1010 kHz	Tooele, Utah
KTNQ	1020 kHz	Los Angeles, California
KDKA	1020 kHz	Pittsburgh, Pennsylvania
KMMQ	1020 kHz	Plattsmouth, Nebraska
KCKN	1020 kHz	Roswell, New Mexico
WBZ	1030 kHz	Boston, Massachusetts
KTWO	1030 kHz	Casper, Wyoming
KCTA	1030 kHz	Corpus Christi, Texas
WWGB	1030 kHz	Indian Head, Maryland
WCTS	1030 kHz	Maplewood, Minnesota
WGSF	1030 kHz	Memphis, Tennessee
KDUN	1030 kHz	Reedsport, Oregon
WDRU	1030 kHz	Wake Forest, North Carolina
WPBS	1040 kHz	Atlanta, Georgia
WHO	1040 kHz	Des Moines, Iowa
WEPN	1050 kHz	New York
KTCT	1050 kHz	San Mateo, California
KRCN	1060 kHz	Longmont, Colorado
WQOM	1060 kHz	Natick, Massachusetts
WLNO	1060 kHz	New Orleans, Louisiana
KYW	1060 kHz	Philadelphia, Pennsylvania
WKNG	1060 kHZ	Talapoosa, Georgia
WIXC	1060 kHz	Titusville, Florida

WAPI	1070 kHz	Birmingham, Alabama
WNCT	1070 kHz	Greenville, North Carolina
WFNI	1070 kHz	Indianapolis, Indiana
WFLI	1070 kHz	Lookout Mountain, Tennessee
KNX	1070 kHz	Los Angeles, California
WDIA	1070 kHz	Memphis, Tennessee
WCSZ	1070 kHz	Sans Souci, South Carolina
WKAT	1080 kHz	Coral Gables, Florida
KRLD	1080 kHz	Dallas, Texas
WTIC	1080 kHz	Hartford, Connecticut
WWNL	1080 kHz	Pittsburgh, Pennsylvania
KFXX	1080 kHz	Portland, Oregon
KMXA	1090 kHz	Aurora, Colorado
WBAL	1090 kHz	Baltimore, Maryland
KAAY	1090 kHz	Little Rock, Arkansas
KFNQ	1090 kHz	Seattle, Washington
WTAM	1100 kHz	Cleveland, Ohio
WZFG	1100 kHz	Dilworth, Minnesota
KNZZ	1100 kHz	Grand Junction, Colorado
KFNX	1100 kHz	Phoenix, Arizona
KFAX	1100 kHz	San Francisco, California
WBT	1110 kHz	Charlotte, North Carolina
KVTT	1110 kHz	Mineral Wells, Texas
KFAB	1110 kHz	Omaha, Nebraska
KRDC	1110 kHz	Pasadena, California
KPNW	1120 kHz	Eugene, Oregon
KMOX	1120 kHz	Saint Louis, Missouri
WDFN	1130 kHz	Detroit, Michigan
WISN	1130 kHz	Milwaukee, Wisconsin
KTLK	1130 kHz	Minneapolis, Minnesota
WBBR	1130 kHz	New York
KWKH	1130 kHz	Shreveport, Louisiana
WQBA	1140 kHz	Miami, Florida
WRVA	1140 kHz	Richmond, Virginia
KHTK	1140 kHz	Sacramento, California

50,000 WATT AM STATIONS

KEIB	1150 kHz	Los Angeles, California
WYLL	1160 kHz	Chicago, Illinois
WCRT	1160 kHz	Donelson, Tennessee
KSL	1160 kHz	Salt Lake City, Utah
KJNP	1170 kHz	North Pole, Alaska
KCBQ	1170 kHz	San Diego, California
KLOK	1170 kHz	San Jose, California
KFAQ	1170 kHz	Tulsa, Oklahoma
WWVA	1170 kHz	Wheeling, West Virginia
KGOL	1180 kHz	Humble, Texas
KOFI	1180 kHz	Kalispell, Montana
WJNT	1180 kHz	Pearl, Mississippi
WHAM	1180 kHz	Rochester, New York
KYES	1180 kHz	Rockville, Minnesota
KERN	1180 kHz	Wasco-Green Acres, California
KFXR	1190 kHz	Dallas, Texas
WOWO	1190 kHz	Fort Wayne, Indiana
WCRW	1190 kHz	Leesburg, Virginia
KEX	1190 kHz	Portland, Oregon
WXKS	1200 kHz	Newton, Massachusetts
WAXA	1200 kHz	Pine Island Center, Florida
WOAI	1200 kHz	San Antonio, Texas
WMUZ	1200 kHz	Taylor, Michigan
KFNW	1200 kHz	West Fargo, North Dakota
WJNL	1210 kHz	Kingsley, Michigan
WPHT	1210 kHz	Philadelphia, Pennsylvania
WHKW	1220 kHz	Cleveland, Ohio
WXYT	1270 kHz	Detroit, Michigan
KFLC	1270 kHz	Fort Worth, Texas
WADO	1280 kHz	New York, New York
WNQM	1300 kHz	Nashville, Tennessee
WJNJ	1320 kHz	Jacksonville, Florida
KPXQ	1360 kHz	Glendale, Arizona
KMNY	1360 kHz	Hurst, Texas
WQLL	1370 kHz	Pikesville, Maryland

KRKO	1380 kHz	Everett, Washington
KZQZ	1430 kHz	St. Louis, Missouri
KTNO	1440 kHz	University Park, Texas
WWNN	1470 kHz	Pompano Beach, Florida
WLQV	1500 kHz	Detroit, Michigan
KSTP	1500 kHz	Saint Paul, Minnesota
WFED	1500 kHz	Washington, D.C.
WMEX	1510 kHz	Boston, Massachusetts
WWBC	1510 kHz	Cocoa, Florida
WLAC	1510 kHz	Nashville, Tennessee
KGA	1510 kHz	Spokane, Washington
WWKB	1520 kHz	Buffalo, New York
KOKC	1520 kHz	Oklahoma City, Oklahoma
KQRR	1520 kHz	Oregon City, Oregon
KKXA	1520 kHz	Snohomish, Washington
WCKY	1530 kHz	Cincinnati, Ohio
KGBT	1530 kHz	Harlingen, Texas
WYMM	1530 kHz	Jacksonville, Florida
KFBK	1530 kHz	Sacramento, California
WDCD	1540 kHz	Albany, New York
KMPC	1540 kHz	Los Angeles, California
WNWR	1540 kHz	Philadelphia, Pennsylvania
KXEL	1540 kHz	Waterloo, Iowa
KRPI	1550 kHz	Ferndale, Washington
WLOR	1550 kHz	Huntsville, Alabama
WAZX	1550 kHz	Smyrna, Georgia
KUAZ	1550 kHz	Tucson, Arizona
KKOV	1550 kHz	Vancouver, Washington
WFME	1560 kHz	New York
WJFK	1580 kHz	Morningside, Maryland
KQFN	1580 kHz	Phoenix, Arizona
KBLA	1580 kHz	Santa Monica, California
WMQM	1600 kHz	Lakeland, Tennessee

APPENDIX E – GMRS/FRS FREQUENCIES

Frequencies are in megahertz (MHz).

Channel	Type	Frequency	FRS Power	GMRS Power
1	FRS/GMRS	462.5625	2W	5W
2	FRS/GMRS	462.5875	2W	5W
3	FRS/GMRS	462.6125	2W	5W
4	FRS/GMRS	462.6375	2W	5W
5	FRS/GMRS	462.6625	2W	5W
6	FRS/GMRS	462.6875	2W	5W
7	FRS/GMRS	462.7125	2W	5W
8	FRS/GMRS	467.5625	0.5W	0.5W
9	FRS/GMRS	467.5875	0.5W	0.5W
10	FRS/GMRS	467.6125	0.5W	0.5W
11	FRS/GMRS	467.6375	0.5W	0.5W
12	FRS/GMRS	467.6625	0.5W	0.5W
13	FRS/GMRS	467.6875	0.5W	0.5W
14	FRS/GMRS	467.7125	0.5W	0.5W
15	FRS/GMRS	462.5500	2W	50W
16	FRS/GMRS	462.5750	2W	50W
17	FRS/GMRS	462.6000	2W	50W
18	FRS/GMRS	462.6250	2W	50W
19	FRS/GMRS	462.6500	2W	50W
20	FRS/GMRS	462.6750	2W	50W
21	FRS/GMRS	462.7000	2W	50W
22	FRS/GMRS	462.7250	2W	50W
*RPT15	GMRS	467.5500	-	50W
RPT16	GMRS	467.5750	-	50W
RPT17	GMRS	467.6000	-	50W
RPT18	GMRS	467.6250	-	50W
RPT19	GMRS	467.6500	-	50W
RPT20	GMRS	467.6750	-	50W
PRT21	GMRS	467.7000	-	50W
RPT22	GMRS	467.7250	-	50W

*Channels 15 – 22 are for repeaters

APPENDIX F - PREPPER RADIO FREQUENCIES

Description	Frequency MHz	Mode*	Tone**
Ham 80 Mtr	3.818	LSB	
Ham 60 Mtr	5.357	USB	
Marine Emer	6.215	USB	
Ham 40 Mtr	7.242	LSB	
Ham 20 Mtr	14.242	USB	
Ham Maritime	14.300	USB	
CB CH 3	26.985	AM	
CB CH 4	27.005	AM	
CB CH 19	27.185	AM	
CB CH 36	27.365	USB	
CB CH 27	27.375	USB	
Ham 10 Mtr	28.305	USB	
Aircraft Emergency	121.500	AM	
Ham 2 Mtr	146.420	FM	100
Ham 2 Mtr Calling	146.520	FM	
Ham 2 Mtr	146.550	FM	151.4
MURS Primary	151.940	NFM***	67
MURS Prepper	154.570	FM	67
Search and Rescue	151.160	FM	127.3
Marine Safety	156.800	FM	

PREPPER FREQUENCIES

Marine Prepper	156.625	FM	
Ham UHF Calling	446.000	FM	100
Ham UHF Prepper	446.030	FM	100
FRS 1 Calling	462.562.5	NFM	67
FRS 2 Chat	462.587.5	NFM	67
FRS 3 Prepper	462.612.5	NFM	67
FRS 4 Prepper	462.637.5	NFM	67
GMRS Ch 17 Prepper	462.600		67
GMS Ch 20 Travel	462.675		141.3

* USB = Upper Sideband, LSB = Lower Sideband
** A tone may not always be required.
*** Narrowband FM

INDEX

Made in the USA
Coppell, TX
04 September 2020

35953944R00085